QUANTITATIVE METHODS FOR PORTFOLIO ANALYSIS

THEORY AND DECISION LIBRARY

General Editors: W. Leinfellner (*Vienna*) and G. Eberlein (*Munich*)

Series A: Philosophy and Methodology of the Social Sciences

Series B: Mathematical and Statistical Methods

Series C: Game Theory, Mathematical Programming and Operations Research

Series D: System Theory, Knowledge Engineering and Problem Solving

SERIES B: MATHEMATICAL AND STATISTICAL METHODS

VOLUME 23

Scope: The series focuses on the application of methods and ideas of logic, mathematics and statistics to the social sciences. In particular, formal treatment of social phenomena, the analysis of decision making, information theory and problems of inference will be central themes of this part of the library. Besides theoretical results, empirical investigations and the testing of theoretical models of real world problems will be subjects of interest. In addition to emphasizing interdisciplinary communication, the series will seek to support the rapid dissemination of recent results.

The titles published in this series are listed at the end of this volume.

QUANTITATIVE METHODS FOR FOR PORTFOLIO ANALYSIS

MTV Model Approach

by

TAKEAKI KARIYA

The Institute of Economic Research,
Hitotsubashi University, Tokyo, Japan

SPRINGER-SCIENCE+BUSINESS MEDIA, B.V.

Library of Congress Cataloging-in-Publication Data

Kariya, Takeaki.
   Quantitative methods for portfolio analysis : MTV model approach /
by Takeaki Kariya.
      p.   cm. -- (Theory and decision library. Series B,
Mathematical and statistical methods ; v. 23)
   Includes bibliographical references (p.   ) and index.
   ISBN 978-0-7923-2254-2   ISBN 978-94-011-1721-0 (eBook)
   DOI 10.1007/978-94-011-1721-0
   1. Portfolio management--Mathematical models. 2. Investment
analysis--Mathematical models. 3. Time-series analysis.   I. Title.
II. Series.
HG4529.5.K37   1993
332.6'01'51--dc20   93-7416

ISBN 978-0-7923-2254-2

Printed on acid-free paper

CONTENTS

PREFACE

This book aims to provide practically useful models and methods for quantitative analysis of financial asset prices, construction of various portfolios, and computer-assisted trading systems. In particular, this book will be beneficial for

1) "quants" (quantitatively-inclined analysts) in financial industries,
2) financial engineers in investment banks, securities companies, derivative-trading companies, software houses, etc., who are developing portfolio trading systems,
3) graduate students and specialists in the areas of finance, business, economics, statistics, financial engineering, and
4) investors who are interested in Japanese financial markets.

Throughout the book a great emphasis is put on the practical usefulness of models and methods for investment decision-making, and some examples are demonstrated with practical analyses and models on Japanese financial markets. The statistical procedures treated in this book are somewhat advanced because our financial world is of many complexities. In fact, the time series variations of financial asset prices are of the complicated features;

(a) they are multi-dimensional phenomena,
(b) their variational structure is evolving over time and
(c) profitable structure of variations will be eventually exploited.

Hence to make effective portfolios, it will be important to empirically model the evolving multi-dimensional variations by a proper multivariate time series model and to manage portfolios constructed from the model with proper trading (rebalancing) rules for profitable investment. Therefore our methods and models for analysis of prices, construction of portfolios and management of portfolios are closely associated with various multivariate time series methods in statistics. In particular, the MTV (multivariate time series

variance component) model the author (1987) proposed plays an important role as a basic information-summarizing and result-stabiliz-
ing model in many ways.   In this book we respect the MPT (modern
portfolio theory) to the extent that the models derived from the MPT
perform well empirically as statistical models for investment
decision making.   In other words, from our viewpoint of practical
usefulness we consider it more important to extract by statistically
scientific methods significant and useful information for investments from available data than to follow the MPT or traditional
finance theory.   The relationship between our approach and the MPT
approach is fully discussed in Chapter 1.   The book consists of the
following three parts:

Part I   Quantitative models for portfolio analysis,
Part II Quantitative asset-allocation systems,
Part III Statistical approaches to option pricing and bond pricing.

In general, the procedure of making and managing a portfolio
consists of the following five elements:

(1) Selection of investment stances.
(2) Selection of a population of financial assets for investment.
(3) Modelling and predicting future prices of financial assets.
(4) Construction of an optimal portfolio.
(5) Choice of a trading (rebalance) rule for managing the
    portfolio.

The most important element in this procedure will be (3) because
an investment is a commitment to uncertainties in future and a better prediction leads us to a better performance.   In Part I, we
introduce basic and practical models and methods for analyzing and
predicting financial asset prices.   In particular, univariate nonlinear models and multivariate time series models are focussed upon.
In addition, we observe some empirical features of returns on
stocks and exchange rates, which can be taken into account in
selecting or screening assets for a portfolio population in (2) and

forming an optimal portfolio in (4). One of the features is nonlinearity for a time series process of (one-period) returns and it
will provide profitable information in option trading. In Part II,
multivariate time series models introduced in Part I are used to
develop various procedures of making portfolios. Above all, we
propose some new procedures such as the procedures of making predictive MTV market portfolios, MTV index portfolios, MTV canonical
correlation portfolios, predictive classification portfolios, etc.
Also some procedures associated with the APT (arbitrage pricing
theory) and Rosenberg's models are introduced with some practices
in the Japanese financial industries. In Part III, practical option
pricing is discussed, where the variational features observed in
Part I are taken into account. A review and development is also
made on the Black-Scholes type option theory. Some new bond pricing
models are also presented in Part III.

Some parts of this book have been adapted from my Japanese book
"*Portfolio Keiryo Bunseki no Kiso* (Foundation for Quantitative Port-
folio Analysis)" Toyo Keizai-shimposha. Some portions of this book
have been used in the course I taught at the Graduate School of
Business of the University of Chicago while I was a visiting profes-
sor. I would like to express my deep gratitude to professors
George. C. Tiao, Ruey. S. Tsay and Arnold Zellner of the University of
Chicago for their various comments, communications and encourage-
ments. I also thank Professors Stanley Pliska of University of
Illinois at Chicago, Edward George of University of Texas A&M,
Regina Liu, Arthur Cohen, Robert Berk, William Strawderman, and
Hiroki Tsurumi of Rutgers University, Narayan C. Giri and Martin
Bilodeau of University of Montreal, Morris L. Eaton of University
of Minnesota, Robert Wijsman of University of Illinois at Urbana,
Jeff C. Wu of University of Waterloo, Peter Kim of University of
Guelph, Bimal K. Sinha of University of Maryland, Arjun K. Gupta of
Bowling Green University, Yoshihiko Tsukuda of Tohoku University,

and Hajime Takahashi of Hitotsubashi University for communications
and encouragements.   I would like to thank Mr.'s Takehiro Enomoto,
Masahiro Fukuyama, Ryuichiro Hirayama, Kimio Horiguchi, Koji Itoh,
Satoshi Kimura, Hiroaki Maruyama, Minoru Masuda, Norikazu Minoshima,
Masayoshi Mizuno, Tei Moriya, Katsuto Ono, Masao Shigematsu, Hideaki
Suzuki, Masafumi Takahashi, Naoya Tatebe, Hiroshi Tsuda, Yukio
Un-no, and Mikio Yamada for personal communications and encourage-
ments in various occasions, and Ms. Midori Shibano for skilled
typing.   Finally I acknowledge the New Japan Securities Financial
Science Foundation, the Nomura Foundation for Social Sciences, the
Trust Companies Association of Japan and the Postal Life Insurance
Foundation of Japan for some financial support.

To my daughter Emi,

Takeaki KARIYA

PART I QUANTITATIVE MODELS FOR PORTFOLIO ANALYSIS
CHAPTER 1
QUANTITATIVE APPROACH TO ASSET ALLOCATION

1 The aim of this book

This book aims to provide a variety of quantitative methods for financial asset allocation. As our terminology, we often use the word **quants** to mean

financial quantitative analysis and/or its methods

or its adjective form rather than the Wall Street nickname for quantitatively inclined financial analysts, the latter of which will be sometimes referred to as "quants" with quotation mark " " The goal of quants asset allocation will be to create through a trading (rebalance) rule a new portfolio price-process (stochastic time series process) out of a set of various asset price-processes with different variation characteristics. The different variation characteristics make it possible to create different portfolio price-processes according to investment preferences or stances. In fact, if the variational processes of financial asset prices were known to us and should remain unchanged without respect to our investment behaviors, the quants asset allocation problem would become a stochastic control problem with a trading rule as a control variable.

However, not only the variational processes of asset prices are unknown to us, but also those variations themselves are the consequences of our behaviors. Hence when we change our behaviors, the processes will change. In fact, financial time series are of the following features;

(a) they are multidimensional phenomena,

(b) economic and financial structure evolves grandually but constantly and

(c) profitable variation structure and opportunities will be eventually exploited, which will in turn change the variation structure of financial time series.

Hence in quants asset allocation, we have to empirically model the changing variation processes via a multivariate time series model and manage portfolios with certain trading (rebalancing) rules. In this procedure, as will be discussed in details in Section 2 and Chapter 6, the most important part will be the third element of the following key elements in **quants asset-allocation system**:

(1) Specification of investment stances on return and risk.

(2) Selection of a target population of assets on which we form a portfolio.

(3) Modelling the multivariate variational process of asset prices in the target population with or without ancillary (fundamental) data.

(4) Construction of an optimal portfolio with respect to a given investment preference on time series performance, where the preference should be associated with some characteristics of the model in (2).

(5) Choice of an optimal trading rule to control the price process of the optimal portfolio with respect to a given investment preference.

In this book we provide some promissing multivariate time series models, which are consistent with some observed features of financial time series, and some statistical techniques to make portfolios, asset allocation and financial products. Also we introduce some practices in quants methods in Japan.

STAT quants is our approach

There will be at least two quants approaches in our world;

(1) MPT (modern portfolio theory) quants approach.

(2) STAT (statistics) quants approach.

In the MPT quants it is considered more important to follow the

traditional finance theory rather than to draw from data significant information for decision-making. On the other hand, in the
STAT quants, which is our approach in this book, it is considered
the most important to analyze available data efficiently and
extract useful information for decision-making. It is noted that
the most relevant information should be on future prices for asset
allocation. Hence we shall treat the MPT in this book within the
scope of this viewpoint. In other words, we require the MPT either
to empirically perform well or to have strong possibilities of
practical implementations. Moreover, the financial time series
phenomena that are of the above features (a), (b) and (c) may not
be well fit in the framework the finance theory assumes. This point
will be discussed in Section 3 in details.

In summary, whether market may be efficient or not, the reason
why we analyze data is to extract by statistical science available
information on future prices contained in data and to make the best
of it for our investments. This is the object of the STAT quants.

2 Demand for variability and for quants portfolio construction system

As is well known, financial markets consist of cash markets in
which interest rates fluctuate, foreign exchange markets as international cash markets, bond and stock markets from which firms get
finance directly, etc. Securities and currencies traded in these
markets are the primary or basic financial instruments whose functions cannot be replaced by the other securities. On the other
hand, derivative securities such as futures and options are
ancillary instruments which promote the *raison d'etre* of the
primary securities (such as optimal allocations of resources) and
the functions of their liquidities, risk-hedge and speculation so
that fair prices be formed. Of course, the prices of these
financial assets are determined by demand and supply. But the

equilibrium prices are not in general macroeconomic equilibrium
prices that equilibriate the real side and financial side of an
economy though they are bound to real side in the long run
through business cycles. As a matter of fact, those prices are
often formed through expectations of various investors who
speculate on their forcasts on future variations of prices to
seek capital gains beyond dividends. Their forecasts are of great
variety, of various sentiments, and hence of great variability.
Speculations on such variable forecasts will shorten holding
periods of stocks or investment period (time horizon) for stocks.
Investors then demand the variability of stock prices and invest
into the variability to seek for capital gains because the income
gains as dividends becomes insignificant relative to capital gains.

The demand for the variability of prices will in turn promote
the tendancy that prices are simply determined by demand and supply
without a strong tie to the real side, and the prices rally in syn-
chronization with expectations. It makes financial prices deviate
from the fundamental values of an economy, increases uncertainties
about future price variations and makes the prediction more difficult. In this situation, the variation due to the logic of a
beauty contest as J.M. Keynes pointed out becomes a big factor for
prediction, and there it will be more important to predict who
majority of people think is the most beautiful. This variational
factor is a *game-theoretic uncertainty* and financial asset price
variations are of this game-theoretic uncertainty as well as of
purely probabilistic uncertainty. This is a notable feature of
financial price variations that are quite different from random
phenomena in natural sciences.

The demand for price variations and the variations of financial
asset prices with game-theoretic uncertainties creates a greater
demand for quantitative methods for analysis and prediction and for
new financial products, instruments and derivative assets. In fact,
we need efficient managements and technologies of processing and

summarizing data and information for timely investment decisions
and we also need efficient technologies of controlling risks in
asset allocations. The development of MPT, mathematical statistics
and computers enables us to pursue such quantitative portfolio
managements and risk control technologies and to make new financial
products which meet the demand to some extent.

Characterization of financial assets and products

   All the prices of financial assets and commodities for investment
such as stocks, bonds, cash, currencies, options, futures, bank
deposits, financial composite commodities, etc. are regarded as
random variables in the sense that their future values are
represented by random variables with specific probability distributions at each future time.   Thereby the differences of these
financial assets and commodities are reflected in the differences
of the forms of the probability distributions, which are in general
unknown to us.   To be more specific, let T be the present time and
$T+k$ the k period ahead time.   Then a financial asset (product)
is represented by a series of random variables representing the
price series

$$\boldsymbol{S} = \{S_t\} = \{S_t: t = \cdots, -2, -1, 0, 1, 2, 3, \cdots, T, T+1, \cdots\}.$$

The past and present values $\mathbb{S}(T-)=\{S_t: t \leq T\}$ of $\boldsymbol{S}$ have been
realized and observed.   And these values are regarded as realized
one by one according to the distribution of each S_t ($t \leq T$) and
the realization of S_t may be affected by its past realized values
S_{t-1}, $S_{t-2}, \cdots$.   On the other hand, the future values $\mathbb{S}(T+)=\{S_t:$
$t > T\}$ of $\boldsymbol{S}$ are not realized yet at $t = T$ and thus unknown.   But
again for each $t > T$, S_t obeys a distribution which commands
which values of S_t are more likely to be realized in future, and
in the realization, S_t will be associated or correlated with other
future S_s's, which will be also affected by the past and present

realized values $S(T-)$. Hence future values of asset prices viewed at T depend on the joint conditional distribution of the future random variables $S(T+)$ given the history $S(T-)$ because the distribution probabilistically governs future time paths of prices. However, no one knows the conditional joint distribution and investors try to figure it out and predict a plausible range of future time paths of S_t. In fact, the present price S_T was realized based on such investors' predictions. In reality, prices of many financial assets are interdependently and simultaneously realized at each time according to the joint distribution of the random variables representing the prices. Hence S_t above must be a vector of different asset prices. In case of a bank deposit with a fixed interest rate, the future values of S_t are realized on the known path in advance with probability 1. But for a longer period there is an uncertainty for change of the interest rate even in the bank deposit.

Consequently the problems of making efficient portfolios and of designing and constructing financial commodities and products are how to control the forms of distributions of future asset prices $S(T+)$ on the basis of investment stances such as risk preferences. The forms of the distributions of $\{ S_t,\ t > T\}$ regulate possible future paths of price realizations probabilistically and give plausible ranges of these paths. To specify the forms of distributions and hence the ranges of paths, the following portfolio-making procedure will be taken.

(1) To make a full use of the past data which includes fundamentals data, such as financial and accounting data of firms etc., as well as the own past realizations of prices $S(T-)$.

(2) To model the price processes of time series variations based on the data via statistical (quantitative) methods, to make a thorough simulation not only with past data but with data being realized, and to check if the processes are approximately grasped by the models to the extent that the models can be used in practice

for decision-making. In this stage the models need not be very
accurate, which in fact is impossible for each asset price series.
But they need to be powerful in describing "large" movements and
tendencies.

(3) To form a composite financial asset (portfolio) based on the
models constructed in (2), and to simulate the process of the composite asset prices to see if the process is a desired one with
respect to a given investment preference. Making a composite asset
will reduce the total risk, and make the stability of model and
increase the controllability of future paths.

It is noted that as environment changes or simply as time goes
by, the distribution of future prices $\mathcal{S}(t+)$ changes with new
information and with changes of people's expectations, and hence
models should be revised or modified or replaced. It is also noted
that at each time t only one observation is realized and hence it
is impossible to know the distribution of S_t at t based on the
single observation. Usually certain inferences are made on the
distribution by investigating some variational features of past
prices. Making inferences into the variational characteristics of
financial asset price processes and providing useful models are the
main themes of Part I, where time series structures are focused
upon.

Investment is prediction

Investments such as financial investments, real investment into
equipments and machines, etc. are in general commitments to future
and need necessarily prediction. Furthermore, the economic behaviors of pursuing profits and avoiding risks in future cannot be
free from prediction. In prediction of financial investments analysis on price variations is the most important and the approaches
customarily used are usually classified as

① fundamentals approach,

② chart (technical) approach, and

③ quants (quantitative, statistical) approach.
In practical predictions these methods may be comprehensively used.
As has been stated, this book treats exclusively quants though
fundamentals data may be used. Of course, the quants may be
connected to the equilibrium analysis in finance theories though
the empirical validity of the theories are not strong, as will be
seen. Further the quants makes it possible not only to capture the
variational structure of prices but also to predict them. The
logic of the prediction is objective, many financial asset prices
are instantly and simultaneously predicted by specific numbers possibly with error probabilities and the prediction errors are also
evaluated. These are some main features of the quants. What is
better, it enables us to construct financial composite commodities
(products) and portfolios with certain variational properties with
respect to an investment preference, in which a Markowitz type portfolio theory may be applied. In practical portfolio managements,
judgemental factors can be combined with the quants if necessary.

Next let us discuss on how to view about prediction in association with the approaches ①, ② and ③. All these three approaches
are common in eventually predicting future movements of prices by
using past and present data. In this sense all the three approaches
are also common in being out of what is called the efficient market
hypothesis in finance theory because they aim to extract the information contained in data for the predictability of prices. In fact,
the efficient market hypothesis denies the profitable predictability. On the other hand, these approaches are different in the ways
of processing available information and predicting future prices.
In the approaches ① and ②, predicted values are not given by
numbers and subjective judgement is a key factor. Hence the
performance of prediction is not explicitly evaluated. Consequently
some people sometimes discuss that the approaches ① and ② are not
committed to prediction, and they tend to deny quants prediction by
numbers. However, all the three approaches exist in order to make

predictions for investment decision-making, and these approaches
are not competitive but complementary.

The quants prediction is often a point prediction and hence it
is often likely to evaluate the validity or effectiveness of the
approach in a simple manner based on posterior small results.
However, an important point in prediction is to understand that we
cannot predict prices precisely as far as the price variations are
random phenomena. That is, the impossibility of predicting future
values of a random variable exactly is nothing but the impossibility
of predicting exactly the value of the sum of dots in throwing two
dice. But in the prediction of the sum of the dots of two dice, it
is optimal to predict it by 7 and there we cannot deny the prediction method even if we observe deviations from 7. In other words,
in the quants prediction we should understand;

a) each prediction does not hit the realized value precisely,

b) in the selection of an objective prediction method, it is
   only possible to select an optimal procedure *on the average.*

3 Modern portfolio thoery (MPT) and quants

Structure of MPT

Following the traditional approach in economics, finance theory
including what is often called modern portfolio theory (MPT) in
general attempts

A. to describe investment behaviors in terms of individual utility-
   maximizing decision making (choice-based theory), and

B. to construct the concept of a general equilibrium by connecting
   financial market with goods (such as consumption goods and real
   capital goods) market (equilibrium theory).

Those theories thus formulated may be valid or effective if our
world is described approximately by the world the theories assume
in the beginning of the development of the logic. However, our
world, and in particular our financial world, seems not only far

from the one described by the initial assumptions but also it is
deviating further and further from it. In fact, in finance
theories, the following assumptions are often made:

(1) Perfect information (foresight).

(2) Perfect competition.

(3) Rational investors (agents).

(4) One-period model (static theory).

(5) Random asset prices.

(6) Homogeneous expectations (forecasts).

(7) Optimization of an objective (utility) function which is
     state-independent.

In such a theory, an investor with perfect information at time 0 is
often assumed to make an investment decision for time 1 at which
not only the investors but also the firms in the theory close their
whole positions and all the profits obtained by the firms are
distributed. There at time 0 the future world is uncertain, and
hence asset prices at time 1 are regarded as random variables. In
other words, the asset prices at time 1 are assumed to obey jointly
a certain probability distribution as probabilistic phenomena. An
investor forms a subjective distribution for the (true) distribution but theories do not allow the investor to have his own
distribution. In fact, most MPT theories assume that investor's
subjective distributions are all common and equal to the true
distribution (an assumption of rational expectations). Consequently, the form of the true population distribution or its characteristics such as means, variances, etc. are known to all the
investors. And under the known distribution, the investors are
supposed to make their investment decisions by opitmizing their
objective functions.

Problems of the MPT

The basic problems of the above theoretical framework are related to the assumptions (1)~(7) on which the theories are constructed. In particular, the following points are crucial when the theories claim their practical or real relevance.

① The distribution or random behavior of asset prices that should be explained by the theories is exogenized.

② The homogeneous expectations on the asset price distribution.

③ The limitation of the one-period model and the market efficiency.

④ The rational behaviors of investors and game-theoretic uncertainty.

In the sequal, we discuss these problems in details.

① **Exogenization of the distribution.** It is common also in quants approach that financial asset prices are regarded as random phenomena with uncertainty. However, the randomness of asset prices is mostly a consequence of the behaviors of investors and it is not exogenously determined. In fact, if the distribution is given independently of investors' decisions and behaviors, an investment decision is a commitment to a random mechanism like a game of throwing two dice. Hence as far as the distribution is assumed to be known,

  a) the differences of the volumes of information (which is in fact excluded by Assumption (1))

  b) the differences of the volumes of fund for investments and

  c) the behaviors of other investors (which is excluded by Assumption (3))

do not affect the asset price distribution at all. Therefore an investment action is a participation to a random game such as a *roulette* in which there is a trade-off between returns and risks.

In such a framework the variations of asset prices are exogenously
and randomly determined without respect to the decisions, behaviors
and demands of investors, and hence the theories give no explanation of the price variations.   Such a framework may be valid in
actuarial sciences, but not in asset pricing theories.

② **Homogeneous rational expectations.**   As has been pointed out, an
investment is a forecast.   The differences in investors' forecasts
create buyers and sellers and equilibrium prices of demand and
supply are a fortiori formed in markets.   The equilibrium prices at
time 1 are random variables at time 0, and as a premise that the
distributions of the equilibrium prices can be formed, the hetero-
geniety of individual's subjective distributions will be necessary.
In fact, if the subjective distributions of investors are common,
the probability that prices rise will be equal to all the people
and hence the differences in people's behaviors will appear only
due to the differences in the objective (utility) functions which
are individually optimized to make investment decisions.   However,
the differences will affect the true distribution of prices because
it does not guarantee the demand and supply equilibrium. Consequent
ly the common subjective distribution will deviate greatly from the
distribution of the equilibrium prices in the market at time 1.
This implies that the common distribution can not be the distribution on which the rational investors make decisions.   It is noted
that the exogenous setting of the distribution of an equilibrium
price together with the homogeneous expectations will strengthen
the aspect of a gambling nature of the problem.   Needless to say,
the assumption of the homogeneous expectations is far from realis-
tic and the basic reason that prices can be bid in markets is that
people form different expectations or forecasts.   Further, the true
distributions of prices are not only unkown but also they may
depend on people's forecasts and behaviors.

③ **The limitation of the one-period model and the adoption of the efficient market hypothesis.** In reality, investments are based on

d) funds of various time horizons for investment.

In other words, the existence of various time preferences in funds also creates fluctuations of asset prices. The one-period model in which an investment is made at time 0 and the economy is closed at time 1 has a serious defect in describing a financial market. The problem becomes more serious when the efficient market hypothesis is combined with the one-period theory for connecting the behaviors of different periods. This hypothesis basically claims that at time t the information accumulated up to $t-1$ is not useful in forecasting future prices and the information available at t is immediately reflected and completely absorbed in prices. Therefore under this hypothesis the probability distribution of prices at $t+1$ depends only on the information at t, and hence arguments can be sectioned one-period by one-period.

In finance, the efficient market hypothesis applied one period by one period to a return process is often associated with the random walk hypothesis of (logarithm of) asset prices. Then returns are determined independently in each period by a random mechanism just like a roulette in which such past information as fundamentals data and own past history makes no effect on the random mechanism. Empirically the random walk hypothesis is strongly rejected.

④ **The rationality of investors and game-theoretic uncertainty.** The assumption of the existence of rational investors has been a target of criticisms for modern economics. But what should be pointed out here is the interdependence of decision makings among investors due to game-theoretic uncertainties. It is difficult to consider the interdependence explicitly in the development of a "theory". In the financial phenomena responding quickly to the

information on other people's behaviors, this aspect will greatly
affect price fluctuations and is a cause of the time series structure of prices.

Exercises

1. State your opinion on the efficient market hypothesis that we get
   no profitable opportunity from publicized information and
   compare it with the arguments of this chapter. What is your
   opinion between the variations of stock prices and business
   cycles?

2. Do your think that the processes of returns evolve when we
   change our investment behaviors or that we change our investment
   behaviors as the variations of returns or prices apparently
   change?

3. Do you share the idea that the processes of returns evolve as
   our economic and noneconomic environments change? Do your think
   that in a financial analysis for investments it is better to use
   data of a longer term?

4. Do you think that in making investment decisions, people use the
   information on price levels?

CHAPTER 2

EMPIRICAL FEATURES OF FINANCIAL RETURNS

1 Introduction

In this chapter in a line with the traditional approach we make some empirical observations on the variational characteristics of financial return series. Such empirical features of variations can be used in asset allocation at least in the following ways.

(1) Screening or filtering of financial assets for investments. In asset allocation we need to select a set of assets from a totality of financial assets to make a portfolio. In practice, accounting variables such as PER (price-earning ratio), retained profit, etc. are often used for selecting such a group by certain criteria. Time series features of variations can be also used as criteria for selecting some assets for portfolios.

(2) Construction of an objective function for portfolios. A variational feature which strongly persists can be an object for constructing a stable portfolio of the feature. If a strong feature of variation is found in some assets and if the feature is desirable relative to an investment preference, a portfolio of the feature can be formed stably and its performance will be predictable. In other words, a predictable portfolio will be possible to make with a time series feature of variations.

(3) Identification of feasible porfolios. It is imoportant to understand the scope of feasible portfolios through some variational features. An objective function which is not consistent with the features is not efficiently optimized and a portfolio obtained through such an objection function will not be stable.

Following this viewpoint, not only variational features of returns but also variational features of any form (function) of prices will be interesting so long as they are profitable and exploitable features in asset allocation. In fact as will be

15

discussed in Chapter 6, one-period return, which is either continuously compound return $x_t = \log S_t - \log S_{t-1}$ or simple return
$x_t = (S_t - S_{t-1})/S_{t-1}$, is not necessarily a convenient form for
quantitative portfolio analysis in asset allocation, where S_t is
an asset price at t. In addition, financial returns $\{x_t\}$ as data
are rather sensitive to noises contained in price levels $\{S_t\}$,
which will be discussed in Section 1 of Chapter 3. It is also
noted that investors use past price levels in addition to past
returns as relevant information in their investment decisions when
they form their forecasts on expected returns and risks.

Apart from these points, we observe the following rather universal variation features of return series of financial assets
such as stocks, currencies, futures on commodities, etc.:

(1) nonnormality, (2) nonindependence and (3) nonlinearity.

Our argument is in a line with Taylor (1986) who observed these
observations tacitly, and we do not use a specific model in this
chapter except for a general time series theory, which we briefly
review below.

Stationary time series

As is well known, a series of random variables, denoted by $\{x_t\}$,
is said to be a stochastic process or a time series process when t
denotes time. Here we assume $t = 0, \pm 1, \pm 2, \cdots$ (discrete time).
For a given $\{x_t\}$, the joint distribution of $(x_{t1}, \cdots, x_{tn})$ is
defined for any time $t_1 < t_2 < \cdots < t_n$ ($n = 1, 2, \cdots$) though it is
unknown in general. In particular the means, variances and autocovariances respectively defined by

(1.3) $\mu_t \equiv E(x_t)$, $\sigma_{tt} \equiv \mathrm{Var}(x_t)$ and $\sigma_{t\,t-k} \equiv \mathrm{Cov}(x_t, x_{t-k})$

($t = 0, \pm 1, \pm 2, \cdots$: $k = 0, \pm 1, \cdots$) are defined through the
distribution when they exist. A (covariance or weakly) stationary
process $\{x_t\}$ is defined as a stochastic process with

(i) $\mu_t = \mu$ for all t and

(ii) $\sigma_{t\,t-k} = \gamma_k$ for all t $(k=0, \pm 1, \cdots)$.

Hence in a stationary process $\{x_t\}$, mean μ, variance γ_0 and autocovariance γ_k for each k are all common for all t. The autocovariance γ_k of x_t and x_{t-k} depends on time difference k, satisfies $\gamma_{-k} = \gamma_k$ by its definition and is estimated by

$$\hat{\gamma}_k = \frac{1}{T}\Sigma_{t=k+1}^{T}(x_t - \overline{x})(x_{t-k} - \overline{x})$$

for given data $(x_1, \cdots, x_T)$. In particular, variance γ_0 is estimated by $\hat{\gamma}_0$. Since the variance γ_0 is constant in a stationary time series, the time series structure of $\{x_t\}$ is described by autocorrelations defined by

$$\rho_k = \gamma_k/(\gamma_0 \cdot \gamma_0)^{1/2} = \gamma_k/\gamma_0 \quad (k=0, \pm 1, \pm 2, \cdots).$$

The set of ρ_k's is called a correlogram, and ρ_k's are estimated by

$$\hat{\rho}_k = \hat{\gamma}_k/\hat{\gamma}_0.$$

Thus under the stationarity of $\{x_t\}$ we can estimate the common mean μ, variance γ_0 and autocorrelations ρ_k in a manner of descriptive statistics. We assume $\Sigma_k|\rho_k| < \infty$. In Chapter 13, the case with $\Sigma|\rho_k| = \infty$ is discussed.

Furthermore the stationarity provides a conceptual and practical facility in modelling a time series $\{x_t\}$ and predicting future values of x. In the Box-Jenkins method, the ARMA (autoregressive moving average) family of stationary models is adopted in their modelling procedure because a stationary process $\{x_t\}$ is approximated as a simpler form by a member of the family. In their procedure the time series structures of $\{x_t\}$ revealed though sample autocorrelations $\hat{\rho}_k$'s are extensively used for model identification.

However, as will be discussed in Chapter 3, the stationarity may be regarded as a relative concept in modelling an economic time series, in particular a financial time series. In fact, a financial time series is of the following characteristics.

1) Economic time series phenomena are multidimensional phenomena, and for a given series, say $\{x_{1t}\}$, there are in general many other series $\{x_{jt}\}$'s which are not only cross-sectionally correlated with x_{1t}, i.e., $\text{Correl}(x_{1t}, x_{jt}) \neq 0$, but also cross-serially correlated with x_{1t}, i.e., $\text{Correl}(x_{1t}, x_{js}) \neq 0$ $(s \neq t)$. And we cannot observe $\{x_{1t}\}$ by deleting the effects of the other variables x_{jt}'s.

2) An economic structure evolves gradually but constantly. This will imply that the autocorrelations of x_{1t}, cross-sectional correlations and cross-serial correlations evolve along with time. In this sense an economic time series may not be stationary.

3) Profitable variation structures may be exploited and may not persist for a long period, where game-theoretic uncertainty may be involved. Investors collect information to exploit profitable opportunities and change their behaviors. Hence the cross-sectional correlations and cross-serial correlations and hence autocorrelations will change by these profitable-opportunity-exploiting behaviors.

Therefore we need a careful procedure in modelling a univariate time series and predicting future values. In particular, characteristic 3) will be strong in financial series and hence a large sample on one variable will not reflect the information on an invariant system over the period. In fact, it is important to select a proper sample period relative to the object of analysis or to the fluctuations we may aim to extract. This will in turn imply that such a modelling method as the Box=Jenkins method which requires a large sample of data needs to be carefully applied. Also note

that a univariate financial time series is quickly synchronized
with many other financial time series. For example, stock prices
and bond prices will be affected by exchange rates and investors'
forecasts, which will in turn change exchange rates through capital
movements among international markets.

2 Random walk and nonnormality
Random walk and its implication

As is discussed in Section 1, it is often assumed in finance
theory that investment decisions are made based soley on returns
$\{ x_t \}$ and independent of price levels $\{ S_t \}$, where

(2.1) $x_t = \log S_t - \log S_{t-1}$.

Further, in finance theory including option theory it is usually
assumed from a viewpoint of an efficient market hypothesis that
return process $\{ x_t \}$ obeys an iid normal (Gaussian) process or an
iid process. Here "iid" means that x_t's as random variables are
independently and identically distributed. This iid property gives
a ground for studies on the common distribution of x_t's based on
time series observations (data) of x_t's ($t=1, \cdots, T$). As a matter
of fact, a great deal of empirical studies have been made on the
intertemporally common distribution of returns of an asset. However,
strictly speaking, such studies are effective only if returns $\{ x_t \}$
are iid. The assumption of the iid-ness of returns is equivalent
to the assumption of a random walk (RW) of log-prices $\{ \log S_t \}$.
The Black-Scholes option theory is based on the Brownian motion of
$\{ \log S_t \}$ as the continuous version of the normal random walk. To
make clear our argument in the sequal, we make

Definition 2.1. The logarithmic price process $\{ \log S_t \}$ is said to
follow a random walk (RW) if the return process $\{ x_t \}$ follows one
of the following processes:

(i)   [Normal Strong RW] $\{x_t\}$ is an iid normal process,

(ii) [Strong RW] $\{x_t\}$ is an iid process,

(iii) [Semi-Strong RW] $\{x_t\}$ is an independent process ,

(iv) [Weak RW] $\{x_t\}$ is an uncorrelated process, i.e.,

$$\text{Correl}(x_t, \ x_s) = 0 \qquad (t \ne s).$$

If $\{\log S_t\}$ is a RW in the sense of one of (i)~(iii), it follows
from the independence that the present and past values of x are
not associated with future returns and hence they have no predictability for future values of x.   In other words, the independence
of x_t's basically denies the profitability by the use of data and
it implies an efficiency of market.   Further it also implies that
univariate time series methods and technical (chart) methods will
lose their analytical power because the methods aim to extract the
time series associations and patterns of x_t's.   Fundamentals approach will not be effective either.   Some more specific remarks
are as follows.

(1) The normal strong RW assumption most gets along with financial
    theory assuming utility functions of investors such as the CAPM
    (capital asset pricing model) and its time-continuous version
    enables us to have the Black-Sholes option pricing formula.
    However, as has been observed in many researches and as is demon-
    strated below, the return processes of most financial assets are
    not normal even if x_t's are iid.

(2) The strong RW assumption is often made for the study on the
    common distribution of x_t's.   A great deal of the past empirical
    researches find out that the distribution is not normal but lepto
    kurtic (thicker-tailed).   As an alternative distribution, normal
    mixture distribution, t-distribution, stable law distribution
    etc. are proposed.   However, the iid-ness of x_t's is denied in
    almost all financial series and hence the strong RW does not hold
    generally (see Taylor (1986) and the examples below).   Contrary
    to the univariate case, not much research on the multivariate

distribution of asset returns has been done. Press (1973) discusses an estimation procedure of multivariate stable law distribution as a possible distribution of multivariate returns. Recently elliptically symmetric distributions are paid some attentions to.

(3) The semi-strong RW is not treated in the literature and it will not hold in practice since a time series structure is found in data.

(4) In the framework of (i) and (ii), the problem of testing the RW hypothesis is also a hot topic in the academic literature and the unit-root tests such as Dickey-Fuller test (1974) and the tests of independence are commonly applied.

However, most financial returns are not independent and hence the distributional results and the RW hypothesis testing results assuming the independence are not valid as they stand.

Descriptive statistics

A first step to modelling financial time series is to look at data and summarize some basic structure and features of the fluctuations by simple "descriptive" statistics. In univariate analysis, most commonly

1) an analysis by basic statistics such as mean ($\bar{x}$), t-value, standard deviation (s), skewness (b), kurtosis (k), etc., and
2) an analysis by sample autocorrelations

are made to extract some variational features from data. However, since we do not know much about the variational structure of the time series process, we do not assume any specific model to look at the data at this stage. Rather, our argument is conditional and tentative. For example, for a sample mean $\bar{x}$ to be meaningful, x_t's need to have a common population mean μ, or at least the mean $\mu_t = E(x_t)$ of x_t cannot change greatly over time. Hence we make a conditional inference on the population mean μ by making a simple assumption on the population structure from which the data are generated, such as the constancy of μ. Then the assumption is checked by a testing procedure, which is often called a diagnostic checking. Taylor (1986) also adopted this viewpoint of empirical modelling and found out tacitly the following universal variation features of financial time series of daily returns; (1) nonnormality (leptokurtosis), (2) nonindependence, and (3) nonlinearity. It is noted that nonlinearity (3) implies (1) and (2). The procedure he took consists of the three steps;

(i)   testing the hypothesis that x_t's are iid normal ($\{\log S_t\}$ is a normal strong RW), the normality is rejected,

(ii) testing the hypothesis that x_t's are iid ($\{\log S_t\}$ is a strong RW), the independence of x_t's is rejected, and

(iii) testing the hypothesis that x_t's are linear (see Section 3 for the definition), the linearity is rejected.

The leptokurtosis of financial series has been discussed by many authors. The nonindependence does not exclude the possibility that x_t's are uncorrelated ($\{\log S_t\}$ is a weak RW). In the next chapter, some nonlinear models which are consistent with the features are considered.

Examples: Nonnormality

A. Exchange Rates

The following table is a summary of the basic statistics computed from the returns of daily Yen/Dollar exchange rates $\{x_t\}$ for 1984.10.9~1987.10.8.

T	$10^5\,\overline{x}$	$10^4\,s$	t	b	k	2*	3*	4*
745	−7.13	6.8	−2.85	−0.90	8.74	39	12	6

  * the number of samples exceeding $\overline{x} \pm j\,s$ ($j=2,3,4$),
     t: t-value.

Some observations from this table and Kariya and Matsue (1989) are as follows.

(i) Sample mean $\bar{x}$ is very small relative to sample standard deviation s and may be negligible for modelling the process of $\{x_t\}$.

(ⅱ) Standard deviations are rather stable even if the sample period is replaced.

(ⅲ) If $\{x_t\}$ is iid normal, the skewness statistic b is not significant.

(ⅳ) If $\{x_t\}$ is iid normal, the kurtosis statistic k is very significant.

(v) The number of samples exceeding $|\bar{x} \pm j\,s|$ is greater than its expected value ($j=2,3,4$) under normality.

From (ⅳ) and (v), the iid normal property of $\{x_t\}$ is rejected and from (ⅲ) the symmetry of the distribution of x_t's is not rejected provided x_t's are iid. In (ⅲ) and (ⅳ) we used the following asymptotic formula for T large: if x_t's are iid normal and T is large,

$$(2.2) \qquad \sqrt{T}\,b \sim N(0,\ 6), \quad \text{and} \quad \sqrt{T}(k-3) \sim N(0,\ 24),$$

(see Kendall and Stuart (1960) for the proof). It is noted that the assumption of normality is necessary for the formula (2.2). Hence in (ⅳ) the rejection of the hypothesis that $\kappa = 3$ (population kurtosis under normality) will lead us to nonnormality. In (v), the expected sample size going out of the interval $(\bar{x} - j\,s,\ \bar{x} + j\,s)$ can be computed by assuming that x_t's are iid normal from $N(\bar{x},\ s^2)$ since T is large. Hence the expected sample size is less than $T \times 0.05 \fallingdotseq 35$ for $j = 2$, less than $T \times 0.005 \fallingdotseq$ 4 for $j = 3$, and less than $T \times 0.0001 \fallingdotseq 0.7$ for $j = 4$. The t-value for μ is significant.

B. Stock Prices (NIKKEI Index, Hitachi)

The next table summarizes the basic statistics of daily returns of the NIKKEI 225 Index and Hitachi for 1983.1.1~1987.12.31.

	T	$10^4\,\bar{x}$	$10^4\,s$	t	b	k
NK 225	1410	7.01	94	2.8	-3.41	72.5
Hitachi	1410	2.39	199	0.5	0.04	7.71

Some observations from this table are as follows.

(i) s is large relative to $\bar{x}$. In particular, the standard deviation s of Hitachi is large (a characteristic of individual stock series).

(ii) If $\{x_t\}$ is iid, the $\bar{x}$ of the Index is significant since $|t|>2$, but that of Hitachi is not significant.

(iii) If $\{x_t\}$ is iid normal, the skewness b of the Index is negatively significant but Hitachi's b is not significant (apply (2.6)).

(iv) If $\{x_t\}$ is iid normal, the kurtosis k's of the Index and Hitachi are both very significant. But the k of the Index is enormous.

From (iv), neither of the Index process and the Hitachi process is iid normal and from (iii), if $\{x_t\}$ is iid, the common distribution of x_t's is strongly skewed to the right for the Index but it is symmetric for Hitachi. Further from (ii), the mean $\bar{x}$, which is sometimes called risk premium, is likely to be positive for the Index. The features observed for Hitachi above are rather universally common to many individual stock series. (See Kariya, Tsukuda and Maru (1990).)

The standard deviations in stock series are observed to be not stable over time.

3 Nonindependent and nonlinear features of a return process
Nonindependence

Define the sample autocorrelation (coefficient) of lag τ by

(3.1) $\quad r_\tau(x) = \sum_{t=\tau+1}^{T}(x_t - \bar{x})(x_{t-\tau} - \bar{x}) / \sum_{t=1}^{T}(x_t - \bar{x})^2$

where $\tau = 1, 2, \cdots$. Of course, for each given τ $r_\tau(x)$ is a form
of the correlation coefficient based on $(x_t, x_{t-\tau})$, $\tau = t+1$,
$\cdots$, T and it measures the degree of linearity between x_t and $x_{t-\tau}$
for $t = \tau + 1, \cdots, T$. To claim the nonindependence of a return
process $\{x_t\}$, we use

Lemma 3.1. (Anderson and Walker (1964)). If a stochastic process
$\{w_t\}$ is iid, the distribution of the sample autocorrelation
$r_\tau(w)$ based on $\{w_t\}$ is approximated by normal distribution
$N(0, \frac{1}{T})$ when T is large where τ is fixed.

If $\{x_t\}$ is iid, then the absolute return process $\{|x_t|\}$ and
squared return process $\{x_t^2\}$ are iid. Hence by this lemma, when
$\{x_t\}$ is iid and T is large,

(3.2) $\quad \sqrt{T}r_\tau(x) \sim N(0, 1), \quad \sqrt{T}r_\tau(|x|) \sim N(0, 1)$, and

$\quad\quad \sqrt{T}r_\tau(x^2) \sim N(0, 1)$.

Applying this result to daily financial return series will usually
reject the hypothesis that $\{x_t\}$ is iid (see Taylor (1986) for the
US series).

A. Exchange Rates

In Figure 3-1 the values of $r_\tau(x)$, $r_\tau(|x|)$ and $r_\tau(x^2)$ are
graphed for lag $\tau = 1, 2, \cdots, 30$ where the sample period is 1984.10.9
$\sim$1987.10.8 (3 years). From this figure it is observed:

(i) $r_\tau(x)$'s are all small and almost all of them are in the 95%
confidence interval $(-1.96/\sqrt{T}, 1.96/\sqrt{T})$ when $T=745$.

(ii) The first lag autocorrelation $r_1(x)$ is very small.

(iii) The first several autocorrelations $r_\tau(|x|)$'s are very significant, which implies that $\{x_t\}$ is not iid.

(iv) For $1 \leq \tau \leq 23$, $r_\tau(|x|)$'s are all positive, which implies together with (iii) that a big variation is likely to be followed by successive big variations.

(v) A similar observation can be made for $r_\tau(x^2)$.

The fact (iii) implies that $\{x_t\}$ is a nonindependent process or $\{x_t\}$ is not an identically distributed process. However, since the variances of x_t's are considered uniformly bounded, if x_t's are independent, the result in (3.2) will hold at least approximately via Lindeberg-Feller Central Limit Theorem. Hence it will be more natural to regard the fact (iii) as an evidence that x_t's are not independent.

Fig. 3.1, 3.2 Graphs of $r_\tau(x)$ (solid line), $r_\tau(|x|)$ (dotted line) and $r_\tau(x^2)$ (broken line)

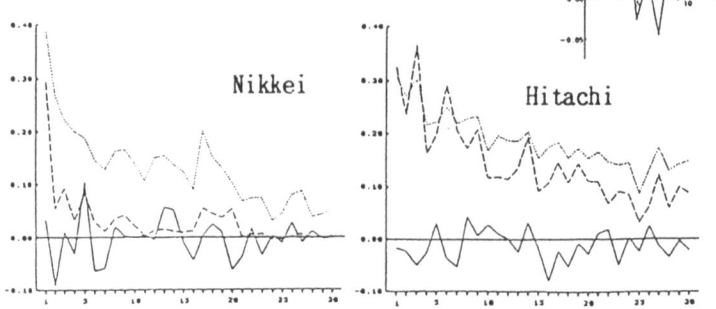

Fig. 3.1 Exchange rate

Nikkei

Hitachi

B. Stocks

In Figure 3-2, the autocorrelations $r_\tau(x)$, $r_\tau(|x|)$ and $r_\tau(x^2)$ of the NIKKEI Index and Hitachi are graphed against $\tau = 1, 2, \cdots, 30$. This figure will indicate:

(i) The $r_\tau(x)$'s of both the Index and Hitachi are all small and almost all of them are in the 95% confidence interval $(-0.052, 0.052)$.

(ii) $r_\tau(|x|)$'s are almost all very significant for both the Index and Hitachi, implying that $\{x_t\}$ is not iid.

(ⅲ) All the $r_\tau(|x|)$'s are positive for the both series, implying that successive big variations are likely to follow a big variation.

(ⅳ) A similar observation is drawn for $r_\tau(x^2)$'s.

The above result holds for almost all individual Japanese stocks (see Kariya, Tsukuda and Maru (1990)).

Nonlinearity

We shall show that our series $\{x_t\}$ is a nonlinear process in the following sense.

Definition 3.1.   A stochastic process $\{x_t\}$ is said to be linear if there exist an *iid* process $\{\varepsilon_t\}$ with mean 0 and variance σ^2, a series of coefficients $\{b_j\}$ with $\sum_{j=0}^{\infty} b_j^2 < \infty$ and a series of means $\{\mu_t\}$ such that x_t is expressed as

$$(3.3) \quad x_t - \mu_t = \sum_{j=0}^{\infty} b_j \varepsilon_{t-j}.$$

To claim nonlinearity, suppose $\{x_t\}$ is linear so that (3.3) holds. Further we assume that ε_t has the fourth moment. Since x_t's can be assumed to be uniformly bounded, this assumption will be legitimate. Let

$$(3.4) \quad Q_t = (x_t - \mu_t)^2,$$

and let $\rho_{\tau,Q}$ and $\rho_{\tau,x}$ denote the population autocorrelations of lag τ of Q_t and x_t respectively. The following formula is due to Taylor (1986) and its proof is outlined in Problem 6.

$$(3.5) \quad 0 \leq \sum_{\tau=1}^{k} \rho_{\tau,Q} \leq \max\left\{\sum_{\tau=1}^{k} \rho_{\tau,x}^2 , \frac{\theta}{(1-\theta)^2}\right\} \quad \text{with}$$

$$(3.6) \quad \theta = 1 - \frac{MSE(x_t^*)}{MSE(\mu_t)} \quad (0 \leq \theta \leq 1), \quad \text{where}$$

(3.7) $\text{MSE}(a_t) = E(x_t - a_t)^2$ and

(3.8) $x_t^* = \mu_t + \sum_{j=1}^{\infty} b_j \varepsilon_{t-j}.$

It is noted that θ is a measure of the relative efficiency of the simple predictor μ_t of x_t (which ignores the time series structure of x_t in (3.3)) compared to the best predictor x_t^* in (3.8). Not only stocks but also currencies are the objects of investment so that x_t's fluctuate rather information-efficiently and hence it is expected that the predictive efficiencies between x_t^* and μ_t are not greatly different. Consequently $\text{MSE}(\mu_t)$ is not very different from $\text{MSE}(x_t^*)$ and θ is expected to be small. Here the information efficiency we are referring to is the "linear information efficiency" as defined as the conditional expectation given past returns. In other words, people in markets rather quickly exploit the linear information contained in past returns and hence there does not remain much linear information when θ is small.

Conversely speaking, there may be a possibility for the non-linear inefficiency of market relative to nonlinear information. For example, if $\{ x_t - \mu_t \}$ is a stationary autoregressive process of order 1, i.e.,

$$x_t - \mu_t = \gamma (x_{t-1} - \mu_{t-1}) + \varepsilon_t$$

then $b_j = \gamma^j$ and hence $\theta = 1 - (1 - \gamma^2) = \gamma^2$. Therefore θ being small implies γ being small and vice versa. In general, θ being small corresponds to small values of autocorrelations $\rho_{\tau, x}$ of $\{ x_t \}$, which has been observed in many financial returns as well as in the exchange rates and stocks of the above examples. In other words, a financial return series $\{ x_t \}$ is usually linear-informa-tion-efficient. By this reason, it will be safe to assume that the relative efficiency θ in (3.6) is at most 0.3:

(3.9) $0 \leq \theta \leq 0.3$,

which implies that the best predictor x_t^* of x_t which makes a full use of the time series structure (3.3) is better up to 30% than the simple predictor μ_t in terms of the mean square error (MSE).

So far, we have assumed that the process $\{x_t\}$ is linear in the sense of (3.3) and that the market is linear-information-efficient in the sense of (3.9). Under these joint assumptions we shall see that the returns of Yen/Dollar exchange rates and of the stock prices treated in the above examples are nonlinear.

A. Exchange Rates

Kariya and Matsue (1989) extensively studied the daily return process $\{x_t\}$ of Yen/Dollar exchange rates over various periods $(1, 2, 3, 4, 5, 7, 10$ years) and found out that $|r_\tau(x)| \leq 0.11$ for all $\tau = 1, \cdots, 30$ and for all the sample periods more than 1 year. Hence we assume

(3.10) $\rho_{\tau, x} < 0.12$

in which case (3.5) is evaluated as

(3.11) $\Sigma_{\tau=1}^{30} \rho_{\tau, Q} < 0.613$.

Since $\bar{x}$ is very small relative to s, considering Q_t, $(x_t - \bar{x})^2$ and x_t^2 equivalent, we estimate $\Sigma_{\tau=1}^{30} \rho_{\tau, Q}$ by $\Sigma_{\tau=1}^{30} r_\tau(x^2)$. Then it follows that all the estimates $\Sigma_{\tau=1}^{30} r_\tau(x^2)$ exceeds far over the upper bound 0.613. In Fiugre 3-1, the autocorrrelations $r_\tau(x)$, $r_\tau(|x|)$ and $r_\tau(x^2)$ of the sample period 1984.10.9~ 1987.10.8 (3 years) are graphed.

B. Stocks

In Kariya, Tsukuda and Maru (1990), the daily return processes of 204 stocks and 2 indices are similarly analyzed. The estimates

$|r_\tau(x)|$'s of $|\rho_{\tau,x}|$'s are less than 0.1 in almost all stocks.
Hence we may assume (3.10) again. Then the same upper bound in
(3.11) obtains. On the other hand, $\Sigma_{\tau=1}^{30} r_\tau(x^2)$ is 0.957 for the
NIKKEI Index, 4.347 for Hitachi and is far beyond 0.613 in almost
all individual series. Hence it is very likely that $\{x_t\}$ is
nonlinear.

A summary of the Japanese Stock Market

Kariya, Tsukuda and Maru (1990), investigated 204 daily stock
price series contained in the Nikkei 225 Index for testing the RW
hypothesis for the period 1983.1~1987.12. Each series consists of
approximately 1,400 returns $\{x_t\}$. In this section, based on their
results, we summarize some basic features of those return processes
in terms of the statistics considered in this chapter.

i) Most of the means $\bar{x}$'s are in the interval (0, 14×10^{-4})
  [192/204]. Only 21 of the 204 series have formal t values
  greater than 2. All the other $|t|$ values are less than 2.

ii) All the standard deviations s_i's are in the interval (0.0148,
  0.0329), whose range is rather small.

iii) The largest skewness is 1.75 and the smallest one is -0.9.
  Most of the skewness statistics b's are in the interval (-0.5,
  1.5) [197/204]. If x_t's were iid, the hypothesis that the
  distribution of x_t's is symmetric would not be rejected.

iv) The largest kurtosis (k) is 57.29 and the smallest one is 4.79,
  and 117 k's are in the interval (4.76, 10) while 79 k's are
  in [10, 20).

v) More than 80% (172 series) of the 204 series have negative
  first-lag autocorrelations $r_1(x)$'s, which is in contrast to
  the results on the US stocks where most stocks have positive
  $r_1(x)$'s. In the pooled autocorrelations $r_\tau(x)$'s ($\tau=1$,
  ..., 30) in the 204 series, only 12% of them exceed 0.05 in
  absolute values $|r_\tau(x)|$ and only 1% of them exceed 0.1 in

absolute values.

vi) More than 95% of all the autocorrelations $r_\tau(|x|)$'s of
absolute returns pooled for 30 lags and 204 series are positive,
among which the largest one is $r_1(|x|)=0.4252$ for Mitsui
Bank and almost all $r_1(|x|)$'s are greater than 0.25.

vii) The NK 225 Index and TOPIX (Tokyo Stock Price Index) have -3.41
and -2.97 respectively for skewness, which are negatively large
in contrast to the skewness of individual series. The kurtosis
of these indices are 72 and 68 respectively, which are surpri-
singly large compared to those of individual series.

Exercises

1. Observe nonnormality, nonindependence and nonlinearity for a
financial daily series by the method in this chapter.

2. Suppose that returns x_t's in (2.1) are iid normal, where
$E(x_t)=\mu$ and $Var(x_t)=\sigma^2$. Then by Definition 2.1
$\{\log S_t\}$ follows a normal strong RW. Show that $E(S_t)=$
$S_{t-1}\exp(\mu+\frac{1}{2}\sigma^2)$. For given $\{S_s : s\leq t-1\}$, how do you
forecast S_t, $\hat{S}_t=S_{t-1}\exp(\mu+\frac{1}{2}\sigma^2)$ or $\hat{S}_t=S_{t-1}\exp(\mu)$
because $E(\log S_t)=\log S_{t-1}+\mu$?

3. (continuation) When $\{\log S_t\}$ is a normal strong RW, show $S_t=$
$S_0\exp(t\mu+\frac{\sigma^2}{2}Z_t)$ for a certain Z_t with $Z_t\sim N(0, t)$,
and for $T>t$, $S_T=S_t\exp(\tau\mu+\frac{\sigma^2}{2}Z')$ for a certain Z'
with $Z'\sim N(0, \tau)$ where $\tau=T-t$. Find $E(S_T)$ and
$Var(S_T)$ given $\{S_s : s\leq t\}$.

4. Suppose that $\{x_t\}$ follows (2.3) with (2.4) and that $\{\log v_t\}$
follows AR(1): $\log v_t=\rho\log v_{t-1}+\eta_t$ where $\{\eta_t\}$ is iid with
$E(\eta_t)=0$ and $Var(\eta_t)=\sigma_\eta^2$. Find $E(S_t)$ and $Var(S_t)$ for
given $\{S_s : s\leq t-1\}$.

5. Figures 3-1 and 3-2 shows that $\{x_t\}$'s are highly and positively correlated. Discuss the way how you use this information
for investment?

6. To prove (3.5), first show

$$\rho_{\tau,Q} = \frac{(\lambda-3)\sum_{i=0}^{\infty} b_i^2 b_{i+\tau}^2 + 2(\sum_{i=0}^{\infty} b_i b_{i+\tau})^2}{(\lambda-3)\sum_{i=0}^{\infty} b_i^4 + 2(\sum_{i=0}^{\infty} b_i^2)}$$

where $\lambda = E(\varepsilon_t^4)$. Next show

$$\rho_{\tau,Q} = [(\kappa-3)\rho_{\tau,y} + 2\rho_{\tau,x}^2]/(\kappa-1)$$

where $\kappa =$ kurtosis of x_t and $y_t = \sum_{j=0}^{\infty} b_j^2 \varepsilon_{t-j}$. Thirdly prove (3.5). {Hint; $\theta = 1-(\sum b_i^2)^{-1}$, $\sum_{\tau=1}^{\infty}\sum_{i=0}^{\infty} b_i^2 b_{i+\tau}^2 = [(\sum b_i^2)^2 - \sum b_i^4]/2$ and $\sum b_i^4 \geq 1$. Also observe $\sum_{\tau=1}^{\infty}\rho_{\tau,y} \leq [(1-\theta)^{-2}-1]/2 \leq \theta/2(1-\theta)^2$ and hence $\sum \rho_{\tau,Q} = \frac{\kappa-3}{\kappa-1}(\sum \rho_{\tau,y}) + \frac{2}{\kappa-1}(\sum \rho_{\tau,x}).$}

7. Study on the proof of (2.2). Note that (2.2) does not hold without normality assumption.

8. Study on the proof of Lemma 3.1.

9. Show that the nonlinearity of a stationary process implies the nonindependence.

10. Study on the proof that the nonlinearity of a stationary process implies the nonnormality.

CHAPTER 3

UNIVARIATE FINANCIAL TIME SERIES MODELS

1 Modelling financial time series

In Chapter 2, we observed that most financial series of daily returns are of the three notable features; leptokurtosis, non-independence and nonlinearity. Hence these features must be taken into account in modelling daily returns of financial assets. For weekly returns these variational features are weakened as empirical evidences though the features are still observed to some degree. In this chapter we consider some nonlinear models which will be use ful not only for modelling univariate financial time series but also for providing nonlinear structures for the MTV (multivariate time series variance component) model discussed in Chapter 5. This chapter is also connected to Chapters 12 and 13 on option pricing.

The nonlinear models considered in this chapter are

① Taylor model,

② ARCH (autoregressive conditional heteroscedastic) model and its extended models.

These models are consistent with the features of financial time series stated above. In addition, as possible models for financial time series we also briefly introduce

③ TAR (threshold autoregressive) model,

④ STAR (smoothing transition autoregressive) model,

⑤ Bilinear model,

⑥ Amplitude dependent model, etc.

Before describing these models, some basic problems involved in time series model-analysis are discussed, namely problems of stationarity, trend and transformations of variables. It is emphasized that these concepts are relative to the objects of financial analysis.

In time series modelling, whether it may be univariate or multivariate, it will be helpful to distinguish between

   (1)   conditional model specification and

   (2)   marginal model specification.

   The conditional specification (1) is often described as follows. Let $\{x_t\}$ be a univariate time series and let $\Psi_{t-1}=\{x_s: s\leq t-1\}$ be the past values of x. Then a model for $\{x_t\}$ is given by specifying the conditional distribution of x_t given Ψ_{t-1}, which is often assumed to be normal;

$$(1.1) \quad x_t \quad \text{given} \quad \Psi_{t-1} \quad \sim \quad N(\mu(\Psi_{t-1}), \ \sigma^2(\Psi_{t-1})).$$

The remaining problem here is how to specify the dependency of conditional mean $\mu(\Psi_{t-1})$ and conditional variance $\sigma^2(\Psi_{t-1})$ on Ψ_{t-1}. In the well-known AR (autoregressive) model, they are linearly specified as

$$(1.2) \quad \mu(\Psi_{t-1}) = \phi_1 x_{t-1} + \cdots + \phi_a x_{t-a} \quad \text{and}$$

$$(1.3) \quad \sigma^2(\Psi_{t-1}) \equiv \sigma^2$$

so that

$$(1.4) \quad x_t = \phi_1 x_{t-1} + \cdots + \phi_a x_{t-a} + \varepsilon \quad \text{with} \quad \varepsilon_t \sim N(0, \sigma^2).$$

This will be one of the simplest models carrying time series structure. It is noted that the distributional assumption is not required to obtain the AR(a) model of order a in (1.4). On the other hand, it is clear that there are numerous ways of deviations from the linearity in (1.2) and (1.3). In fact, the ARCH model keeps (1.2) for the conditional mean but specifies the conditional variance as

$$(1.5) \quad \sigma^2(\Psi_{t-1}) = \theta_0 + \theta_1 \varepsilon_{t-1}^2 + \cdots + \theta_b \varepsilon_{t-b}^2$$

with ε_t in (1.4), which is nonlinear in Ψ_{t-1}. On the other hand, an amplitude dependent model proposed by Ozaki (1980) specifies the mean as

(1.6) $\mu(\Psi_{t-1}) = \phi_1(\Psi_{t-1}) x_{t-1} + \cdots + \phi_a(\Psi_{t-1}) x_{t-a}$

with the conditional variance kept constant as in (1.3), where

(1.7) $\phi_j(\Psi_{t-1}) = \phi_{j0} + \phi_{j1}\exp(-\gamma x_{t-1}^2) \quad (j=1,\cdots, a)$.

Of course, we can construct a model with mean (1.6) and variance (1.5). In other words, there are various ways to construct a non-linear model and the validity of such a model is given by its analytical power in specific and practical applications, not by its form, though we usualy require an interpretability of nonlinear structure in advance relative to a phenomenon to analyze.

Marginal specification is sometimes considered when the cross-sectional structure of a phenomenon is clear from some empirical observation to some extent but when the time series structure is not clear. An example is Taylor's model for financial returns given by

(1.8) $x_t - \mu = u_t v_t, \quad v_t > 0$

where $\{u_t\}$ and $\{v_t\}$ are independent time series processes, and $u_t \sim N(0, 1)$. But no specific time series structure is assumed for $\{u_t\}$ and $\{v_t\}$ because it is not quite known. In (1.8) v_t^2 is the conditional variance for x_t given v_t but it is *not* the conditional variance given past history Ψ_{t-1} as it stands. In the MTV (multivariate time series variance component) model treated in Chapter 5, this marginal specification approach is taken.

2 Stationarity, trend and return

When a time series model is proposed in the statistical literature, it is often the case that stationarity is a central issue of the literature and some sufficient conditions for stationarity are

provided together with estimation and diagnostic checking procedures including model selection procedures. Here we discuss the
role of stationarity in financial analysis and its implication.

As has been stated in Chapter 2, a financial time series such as
stock prices has a direct and indirect relationship with many other
variables in our economy. The relations among the variables are
not only cross-sectional but also cross-serial. For example, a
change in exchange rates at $t-1$ is likely to affect stock prices
at t and vice versa, so that the variables are mutually inter-
dependent over different periods. However in univariate time
series modelling these cross-sectional and cross-serial relationships among the variables are ignored. In addition, the relations
change over time. In fact, cross-sectional relationship is changing constantly with changes of polices, individual preferences,
exogenous economic and noneconomic shocks, etc. In this sense, a
univariate time series model in economic analysis will be a short-
term model. This aspect seems in particular strong in financial
analysis. Hence sample periods for univariate financial analyses
should not be long. This will be true even if such a model as a
time varying coefficient model is used because the model always
carries some time-invariant parameters.

On the other hand, the stationarity of a stochastic process
$\{x_t\}$ is a strong concept about a time invariant structure of the
process (model). The covariance stationarity of $\{x_t\}$ is defined
as a process satisfying

(i) $E(x_t)=0$ (or $E(x_t)=\mu$), and

(ii) $\mathrm{Cov}(x_t, x_{t-k})\equiv\gamma_k$ is independent of t.

This definition implies that the system which generates data is
invariant in time and independent of historical time t so long as
the mean and autocovariances are concerned. Since the autocorrelation $\rho_k=\gamma_k/\gamma_0$ between x_t and x_{t-k} is equal to the autocorrela
tion ρ_{-k} between x_t and x_{t+k} whatever t may be, the concept of
stationarity is a concept closely related to a series of random

variables continuing from $-\infty$ to ∞. Consequently this concept implies that each observation from the system is not directly dependent on the observation timepoint. However, as we discussed above, financial and economic time series data do depend on their observation timepoints as historical time. In other words, our social and economic system is not time invariant, and the auto-correlation between x_{s-k} and x_s is likely to be different from the one between x_{t-k} and x_t when $|t-s|$ is relatively large. Also the variances of x_t and x_s will be different for $|t-s|$ large.

Nevertheless, we can assume a stationary model for an economic time series by choosing a suitable length of sample period. In fact, we can consider our economic structure a time invariant system for a suitably chosen sample period, and conceptually regard it as a system continuing from $-\infty$ to $+\infty$ and data as generated from it for the sample period. However this does not mean that past data far before the sample period and future data far after the sample period can be regarded as generated from the same economic structure, but that only the data in and near around the properly chosen sample period is regarded as generated from a common structure. From this viewpoint the stationarity given to a model is simply considered a conceptual facility and framework to provide a statistical tractability of model identification, estimation and prediction. A model obtained through such a procedure based on the stationarity will be a short-term model for prediction. In general there will be a trade-off between the predictability and stability of a model. In fact, the longer the sample period is, the more stable a model on it will be, and the less predictive power the model will have. This is because a model based on a longer sample period is most likely to explain the long past fluctuations of data as a whole and to become an averaged and stabilized model over the period. If the system is stable, the longer the sample period is, the more predictive power the model

will have. But in univariate financial time series, this will not
be the case and recent data will carry more information on the near
future. Hence for prediction it will be advisable to select a
shorter sample period to the extent that the stability of model is
not seriously damaged. Thus in such a series, as stated above,
stationarity is simply a conceptional tool and the continuously
smooth change of economic structure will guarantee the smooth
change of cross-sectional and cross-serial correlations and autocorrelations, through which the predictability of a model is
secured for a comparatively short sample period.

Trend and return

Another important problem when a stationary model is applied to
economic time series data is the problem of trend. Because a
series with a trend is not stationary, it is often the case that
the series is detrended. But the concept of trend is rather vague,
there are many ways to detrend a series, and the predictive perform
ances of a model greatly vary with detrending methods. This will
imply that a stationary model-analysis and its prediction is always
subject to the detrending method when data is adjusted for trend,
and hence both the empirical results obtained under a specific
detrended series and the stationarity secured by the detrended
series are rather tendative or conditional. A detrending method
often used is the difference method in which a given series $\{ U_t \}$
is transformed into $\{ x_t \}$ as

(2.1) $x_t = \Delta U_t = U_t - U_{t-1}$ or in general $x_t = \Delta^d U_t$.

In the Box=Jenkins method, the difference method is systematically
incorporated in the ARIMA (autoregressive integrated moving
average) family of models with the concept of stationarity.
However, strictly speaking, an integrated process does not get
along with stationarity. In fact, for example, suppose $\{ x_t \}$ in

(2.1) is an AR(1) process (i.e., $\{U_t\}$ is ARI(1,1));

(2.2) $x_t = \rho\, x_{t-1} + \varepsilon_t$, $|\rho| < 1$ with
              $\{\varepsilon_t\}$ iid and $E(\varepsilon_t^2) < \infty$

so that $\{x_t\}$ is stationary. Then by its definition for any k

(2.3) $U_t = x_t + x_{t-1} + \cdots + x_{t-k} + U_{t-k-1}$
              $= \varepsilon_t + \varepsilon_{t-1} + \cdots + \varepsilon_{t-k} + U_{t-k-1} + \rho\,(\Sigma_{j=1}^{k+1} x_{t-j})$

for any finite k. Here though $x_t = \Sigma_{k=0}^{\infty} \rho^k \varepsilon_{t-k} < \infty$ (in L_2),
U_t diverges as $k \to \infty$. In other words, the stationarity of $\{x_t\}$
implies the finiteness of $\Sigma_{k=0}^{\infty} \rho^k \varepsilon_{t-k}$ and Correl$(x_t, x_{t-k}) =$
ρ^k for any large k, while as $k \to \infty$, U_t diverges in (2.3),
which contradicts with the finiteness of U_t at each t. That is,
the stationarity of $\{x_t\}$ is not well consistent with the actual
finiteness of $\{U_t\}$. But this does not mean that the difference
method is inappropriate nor that the Box=Jenkins is invalid. An
important point is that both the trend and the stationarity are
relative concepts in economic time series to efficiently extract
relevant information contained in data, and we should be flexible
and practical in using these concepts. It will be rather absurd to
drive out relevant information by enforcing stationarity strictly.

   This problem is closely related to financial analysis. In
finance theory it is customarily assumed that investors make their
investment decisions based on expected returns and risks (variances
of returns). Hence in almost all financial analyses, return series
defined by $x_t = \log S_t - \log S_{t-1}$, where $\{S_t\}$ is a price series, is
used. However, people in reality use the price information as well
as the information on returns for their investment decisions, and
when the price levels are higher, people tend to think the corresponding expected returns smaller. Furthermore, as will be discussed in Chapter 6, it is appropriate to analyze the fluctuations of
relative price levels in order to construct portfolios through

quants.   Of course, from

(2.4) $\log S_t = x_t + x_{t-1} + \cdots + x_{t-k} + \log S_{t-k+1}$,

the analysis on $\{x_t\}$ may also provide relevant information on
price levels $\{S_t\}$. However, as has been pointed out in Chapter 2,
the return process $\{x_t\}$ is very likely to be affected by noises.
Let us argue the problem here in details.   Suppose that the log-
price series $\{\log S_t\}$ is contaminated by noises as

(2.5) $\log S_t = A_t + \xi_t$,

where $\{A_t\}$ is a main systematic movement including a stochastic
trend and $\{\xi_t\}$ is a white noise.   Also let the relative contribution of ξ_t to $\log S_t$ be

(2.6) $\eta_1 = \text{Var}(\xi_t)/\text{Var}(\log S_t)$.

Then for the return process $x_t = A_t - A_{t-1} + \xi_t - \xi_{t-1}$, the
relative contribution of $\xi_t - \xi_{t-1}$ to x_t is

(2.7) $\eta_2 \equiv \text{Var}(\xi_t - \xi_{t-1})/\text{Var}(x_t) \doteqdot \eta_1/(1-\lambda)$ with
           $\lambda = \text{Correl}(\log S_t, \log S_{t-1})$

provided $\text{Var}(\log S_t) \doteqdot \text{Var}(\log S_{t-1})$.   If the correlation λ of log
S_t and $\log S_{t-1}$ are positively and highly correlated, which is a
typical case in financial series, then the relative contribution of
$\xi_t - \xi_{t-1}$ to x_t is much greater than that of ξ_t to $\log S_t$.   For
example, if $\eta_1 = 0.05$ and $\lambda = 0.8$, then $\eta_2 = 0.25$, and if $\eta_1 = 0.05$
and $\lambda = 0.9$, then $\eta_2 = 0.5$.   In addition, in the periods for which
$A_t - A_{t-1}$ is small, the fluctuations of x_t are mainly caused by
$\xi_t - \xi_{t-1}$ and move like MA(1) $\xi_t - \xi_{t-1}$ since ξ_t is a white
noise.   Of course this argument is subject to the decomposition (2.
5).   But for example, when $\{S_t\}$ is a stock price series, then (2.5)
is rather likely with say, A_t fundamentals and ξ_t news.   Under

such a situation, the analysis of returns will not efficiently
provide relevant information for investments.

To close this section, I would like to remind the readers that
the object of statistics is to provide other positive sciences
which use data with various concepts and methods to "science" data,
and that statistics cannot exists for its own sake.

3 Taylor model

As a model for a daily financial return process $\{x_t\}$ that is
consistent with the variational features observed in Chapter 2;

(3.1)   (1) leptokurtosis   (2) nonindependence   (3) nonlinearity,

Taylor (1986) proposed the following product model;

(3.2) $x_t - \mu = u_t v_t$ $v_t > 0$,   where

(3.3)   (a) $\{u_t\}$ is a normally stationary process with mean 0
              and variance 1 and

         (b) $\{u_t\}$ and $\{v_t\}$ are independent.

We call (3.2) with (3.3) Taylor model. Under the assumption (3.3),
u_t and v_t are uniquely identifiable and conditional on v_t

(3.4) x_t given $v_t \sim N(\mu, v_t^2)$,

and hence v_t is the conditional standard deviation. This implies
that the marginal distribution of x_t is a normal mixture and therefore the kurtosis of x_t is greater than 3, implying the leptokurtosis. Further, the process of the conditional standard
deviations v_t's is not specified in (3.3) and hence in general v_t
changes with time t. This model is also shown to be consistent
with the observations on the autocorrelations of x, $|x|$ and x^2
in Chapter 2.

In a line with Tauchen and Pitts (1983), a derivation of the model was provided in Taylor (1986).

Specification of $\{v_t\}$

The time series structure of conditional standard deviation v_t is quite important to know the sizes of future variations (volatilities) of returns. In particular, the knowledge of the sizes provides a profitable opportunity for a straddle position in options (see Chapter 12). As processes of $\{v_t\}$, Taylor (1986) considered 1) logAR(1) model, 2) ARMA(1,1) model and 3) exponentially smoothing model. However, a big problem is that because of the unobservability of $\{u_t\}$ and $\{v_t\}$, we cannot develop a complete model-identification procedure. From the relation

$$E(|x_t - \mu|) = (2/\pi)^{1/2} E(v_t),$$

Taylor estimated v_t by

$$(3.5) \qquad \hat{v}_t = |x_t - \bar{x}| / \delta \qquad \text{with} \quad \delta = (2/\pi)^{1/2} = 0.798$$

and checked the applicability of the models. Through least squares fitting, the exponentially smoothing model is recommended as a general volatility process $\{v_t\}$ of financial asset returns.

In practice, as far as $\hat{v}_t$ is regarded as an estimate of v_t, a usual AR model will perform better than the exponentially smoothing model for prediction. Another possible approach will be a signal extraction method applied to

$$\log|x_t - \mu| = \log v_t + \log|u_t|$$

(see Pagano (1977)).

Estimate of u_t and standardized returns

When we predict x_t or when we test the uncorrelatedness of x_t's, we need an estimate of the unobservable u_t. Since u_t is the standardized return

(3.6) $u_t = (x_t - \mu)/v_t,$

using an estimate $\hat{v}_t^*$ of the conditional standard deviation and the sample mean, we may estimate u_t by

(3.7) $y_t = (x_t - \bar{x})/\hat{v}_t^*.$

Here $\hat{v}_t^*$ is different from $\hat{v}_t$ in (3.8). In fact, if $\hat{v}_t^* = \hat{v}_t$, y_t in (3.7) takes only 1 or -1. A reasoning for the definition of y_t in (3.7) is as follows. The conditional standard deviation v_t is regarded as determined by its own process reflecting market activities. Hence if $\{v_t\}$ follows such as an AR(1) model $v_t = \lambda v_{t-1} + \varepsilon_t$, the model value of v_t as a systematic part of v_t is $\hat{v}_t^* = \lambda v_{t-1}$ and hence it will be reasonable to replace v_t in (3.6) by the systematic part $\hat{v}_t^*$ of the model with the error part deleted. This argument will be more sensible if we model $\log v_t$ by an AR model since v_t is a scale factor of x_t. Anyway y_t thus defined is called estimated standardized return in the sequal.

Taylor proposed as the normalized factor $\hat{v}_t^*$ the systematic part of the exponentially smoothing model;

(3.8) $\hat{v}_t^* = (1-\gamma)\hat{v}_{t-1}^* + 1.28\gamma|x_t - \bar{x}|$ $(t \geq 21),$
         $\hat{v}_{20}^* = 1.28\Sigma_{t=1}^{20}|x_t - \bar{x}|/20,$

where γ is estimated so as to minimize

$$\Sigma_{t=21}^{T}(\hat{v}_t - \hat{v}_t^*)^2 \quad \text{with} \quad \hat{v}_t = 1.28|x_t - \bar{x}|.$$

The parameter γ is called smoothing parameter and denotes the rate of the dependence of $\hat{v}_t^*$ on the variation $|x_{t-1} - \bar{x}|$ at $t-1$. If we adopt this framework, the return x_t is expressed as

$$x_t = \bar{x} + y_t\hat{v}_t^*,$$

and if we model $\{y_t\}$ by an AR model for example, future values of x will be predictable since future values of $\hat{v}^*$ are predicted by the model (3.8) through repeated substitutions. An asymptotic validity for the use of y_t for u_t is provided in Taylor (1986).

4   ARCH model

The ARCH (autoregressive conditional heteroscedastic) model Engle (1982) and Engle and Kraft (1983) proposed also takes into account the variations of conditional variances. An ARCH model is of the form

(4.1) z_t given $\Psi_{t-1} \sim N(g_t, h_t)$ with
$$g_t = \phi_1 z_{t-1} + \phi_2 z_{t-2} + \cdots + \phi_p z_{t-p},$$
$$h_t = \delta_0 + \delta_1 \varepsilon_{t-1}^2 + \cdots + \delta_q \varepsilon_{t-q}^2, \quad \text{and}$$
$$\varepsilon_t = z_t - g_t,$$

where $\Psi_{t-1} = \{ z_{t-j} : j \geq 1 \}$ is the past history of z, which is often referred to as the past information of z. A more familiar form is

(4.2) $z_t = \phi_1 z_{t-1} + \cdots + \phi_p z_{t-p} + \varepsilon_t$ with
$$E[\varepsilon_t \mid \Psi_{t-1}] = 0,$$
$$\text{Var}(\varepsilon_t \mid \Psi_{t-1}) = h_t = \delta_0 + \delta_1 \varepsilon_{t-1}^2 + \cdots + \delta_q \varepsilon_{t-q}^2, \quad \text{and}$$
$$\varepsilon_t \text{ given } \Psi_{t-1} \sim N(0, h_t).$$

Hence h_t is the conditional variance of z_t given Ψ_{t-1} and it is a linear function of past squared errors ε_{t-1}^2, $\varepsilon_{t-2}^2, \cdots$. This specification of conditional variance implies that a large variation is likely to be followed by consecutive large variations and small by small, which is consistent with the observation made in Section 2. A basic difference with Taylor's model is that the ARCH model is conditionally specified in the sense of Section 1 and hence the conditional variance h_t is a deterministic function of the past z_{t-j}'s through error terms ε_{t-j}'s. In Taylor's model, the variations of asset returns are regarded as responsive to the

t-th information as well as past information and hence the conditional variances are assumed to follow a process.

To make this point clearer, it is useful to understand the difference between

1) unconditional model specification and

2) conditional model specification

(see Section 1). Let $\{z_t\}$ be a time series or a stochastic process. In 1), without respect to the past information Ψ_{t-1} a model is specified marginally. In Taylor's model $z_t = x_t - \mu_t = \varepsilon_t$ with $\varepsilon_t = u_t v_t$ where $\mu_t = E[x_t]$ is an unconditional mean of x_t at t and $\{u_t\}$ and $\{v_t\}$ are independent processes which can be specified in various ways. The $v_t^2 > 0$ is the conditional variance of x_t but the conditioning is made on its own variable v_t of time t, not on the past variables of x's (nor those of v's). On the other hand, in 2) the conditioning is made on the past information Ψ_{t-1}. In fact, as is shown in (4.1) and (4.2), g_t and h_t are specified as the conditional mean and conditional variance of z_t given the past Ψ_{t-1}. From (4.1), writing

$$(4.3) \qquad u_t = \frac{z_t - g_t}{v_t} = \frac{\varepsilon_t}{v_t} \qquad \text{with} \quad v_t = h_t^{1/2},$$

the ARCH model is also expressed as

$$(4.4) \qquad z_t - g_t = u_t v_t \qquad \text{with} \qquad u_t \sim \text{iid } N(0,1),$$

which is of a similar form as Taylor's model. But here $\{u_t\}$ and $\{v_t\}$ are not independent. In fact, by $\varepsilon_t = u_t v_t$,

$$(4.5) \qquad v_t^2 = h_t = \delta_0 + \delta_1 u_{t-1}^2 v_{t-1}^2 + \cdots + \delta_q u_{t-q}^2 v_{t-q}^2.$$

If $\phi_i = 0$ ($i = 1, \cdots, p$) in (4.1), then $g_t = 0$ in which (4.4) is the same form as Taylor's model, but $\{u_t\}$ and $\{v_t\}$ are still heavily dependent.

The dependence of $\{u_t\}$ and $\{v_t\}$ makes it difficult to study the empirical features of financial asset returns in the model. In fact, in an ARCH model conditions for the stationarity of $\{z_t\}$ have been studied considerably, but the higher order autocorrelations of $\{z_t\}$ and the moments are hard to derive. Further the autocorrelations of $\{|x_t|\}$ and $\{x_t^2\}$ are very difficult to study in the model. Consequently such observations that $r_\tau(x)$'s are small but $r_\tau(|x|)$'s and $r_\tau(x^2)$'s are big are not easily studied in the ARCH model. On the other hand, in the ARCH model the estimation procedure can be based on the standard maximum likelihood (ML) method through a numerical optimization and a general asymptotic theory can be applied to the study of sampling properties of the estimators. A warning is that the numerical optimization will not necessarily give the maximum value of the likelihood function. In addition, the existence of a global maximum is not proved.

Nevertheless, an ARCH model is no doubt a promising model for modelling a financial time series. The model can be modified in many ways. The basic feature of the model is the conditional specification of the variance;

$$h_t = \mathrm{Var}(z_t \mid \Psi_t) = h(\varepsilon_{t-1}, \varepsilon_{t-2}, \cdots\cdots).$$

(1) Nelson (1991) specifies, for example for $h_t = h(\varepsilon_{t-1})$

   i) $h(\varepsilon_{t-1}) = a\exp(b\varepsilon_{t-1})$ because h_t needs to be positive, and

   ii) hence h_t is not necessarily symmetric because $h(-\varepsilon_{t-1}) \neq h(\varepsilon_{t-1})$.

The asymmetry of $h(\cdot)$ is justified because the upward variations of asset returns are not necessarily the same as the downward variations. (See also Taylor (1986).)

(2) The conditional heteroscedasticity can be introduced to a usual regression model;

$$y_t = x_t\beta + \varepsilon_t, \qquad \varepsilon_t \mid \Psi_{t-1} \sim N(0, h_t),$$
$$h_t = \alpha_0 + \alpha_1 \varepsilon_{t-1}^2.$$

(3) ARCH-M model proposed by Engle, Lilien and Robin (1987) is

$$y_t = x_t \beta + \gamma \, h_t + \varepsilon_t.$$

Here the conditional variance serves as an explanatory variable in
a regression model. A motivation behind the model is as follows.
A risk-averse agent demands higher risk premium when a market is
more uncertain, and hence the size h_t of uncertainty in financial
markets (predicted at $t-1$) affects the risk premium (y_t).
(4) Bollerslev (1986) generalizes the ARCH model to what he calls
the GARCH model with the condition variance

$$h_t = \alpha_0 + \Sigma_{i=1}^{q} \delta_i \varepsilon_{t-i}^2 + \Sigma_{j=1}^{k} \gamma_i \, h_{t-j}.$$

(5) Tsay (1987) shows that a random coefficient autoregressive (RCA)
model contains the ARCH model and generalizes the RCA model to the
case where an MA part is included. In the RCA model the conditional variance is expressed in a general quadratic form of ε_{t-j}'s as

$$h_t = \delta_0 + \Sigma_{j=1}^{q} \Sigma_{k=1}^{q} \gamma_{jk} \varepsilon_{t-j} \varepsilon_{t-k}.$$

Applications of ARCH model

A. Exchange rates

Milhoj (1987) applied the ARCH model to the analysis of the
daily returns $\{x_t\}$ of exchange rates (SDR currency basket – a portfolio of currencies). He adopted the proposition of Engle (1982)
that $\phi_i = 0$ ($i=1, \cdots, p$), hence $g_t = 0$ in (4.1) and $q=4$, so
that

$$h_t = \delta_0 + \delta_1 \varepsilon_{t-1}^2 + \cdots + \delta_4 \varepsilon_{t-4}^2.$$

The three periods are treated along the periods of changes of the
composition of the SDR. In the period 1981.1.1~1986.1.1, the composition ratios (portfolio ratios) of the SDR currencies are 42%
(US Dollar), 19% (German Mark), 13% (Sterling Pound), 13% (French

Franc) and 13% (Japanese Yen).   The summary statistics and sample
autocorrelations are as follows.

$\bar{x}$	s	k	r_1	r_2	r_3	r_5	r_6
$-.116\times10^{-3}$	$.115\times10^{-4}$	6.40	$-.03$	.01	.07	$-.02$	.002

Compared to the results in Section 2, this result is of the feature
that $|\bar{x}|$ is larger than standard deviation s (portfolio effect).
The kurtosis $k=6.40$ is very significant if x_t's are iid normal.
If $\{x_t\}$ is iid, $r_3=0.07$ is significant at 5%.   The estimates of
coefficients δ_j's in the conditional variance h_t by the ML method
are

$$\delta_0=.11\times10^{-4},\ \ \delta_1=.045,\ \ \delta_2=.056,\ \ \delta_3=.149,\ \ \delta_4=.249,$$

where the iid normality is assumed for $\{u_t\}$ with u_t in (4.3).   To
check the iid normality, based on the standardized returns

$$\hat{u}_t = x_t / \hat{v}_t \qquad \text{with} \qquad v_t=h_t^{1/2},$$

he computed the kurtosis k, 50% range $f_{.75}-f_{.25}$, 20% range
$f_{.60}-f_{.40}$ and Ljung=Box test $\bar{Q}_{50}$.

k	$f_{.75}-f_{.25}$	$f_{.60}-f_{.40}$	$\bar{Q}_{50}$
5.37	1.31	0.50	58.5

   This result shows that the kurtosis k of $\hat{u}_t$ gets smaller than
that of x_t.   But it is still highly significant if u_t's are iid
normal.   Also the 50% and 20% ranges are both very significant
under iid normality.   Therefore the iid normality is denied.   He
refers the cause of the nonnormality to outliers in data.   While,
the Ljung=Box test $\bar{Q}_{50}$ is not significant and hence it does not
provide an evidence for the serial correlation of $\hat{u}_t$ through this
statistic.   It is noted that the estimates are derived under

normality assumption by the ML method.

B. Stock prices index

Akgiray (1989) analyzed the daily returns $\{x_t\}$ of the Stock Index in the Center for Research in Security Prices (University of Chicago). He confirmed the three main features of variations of returns (nonnormality, nonindependence, nonlinearity) as in Taylor (1986) and then applied the ARCH and GARCH models. Setting $z_t = x_t$ in (4.1), his model is

$$g_t = \phi_0 + \phi_1 x_{t-1}, \quad \varepsilon_t = x_t - g_t, \quad \text{and}$$
$$h_t = \delta_0 + \sum_{i=1}^{q} \delta_i \varepsilon_{t-i}^2 + \sum_{j=1}^{r} \beta_j h_{t-j}.$$

Using the ML method and selecting a model by likelihood ratios, he reported the following results for 1981~1986.

① ARCH model ($q = 2$, $r = 0$)

ϕ_0	ϕ_1	δ_0	δ_1	δ_2	σ_e^2
$.47 \times 10^{-3}$	.1406	$.6352 \times 10^{-4}$	.03419	.01773	$.67 \times 10^{-4}$

Here

$$\sigma_e^2 = \delta_0 / (1 - \sum_{i=1}^{p} \delta_i), \qquad \sigma_x^2 = \sigma_e^2 / (1 - \phi_1^2).$$

He confirmed that the hypothesis $\delta_1 = \delta_2 = 0$ is rejected by Lagrangean test and χ^2 test and that the squared residuals e_t^2's are of autocorrelation structure.

② GARCH ($q = 1$, $r = 1$)

ϕ_0	ϕ_1	δ_0	δ_1	β_1	σ_e^2	σ_x^2
$.49 \times 10^{-3}$	.1405	$.223 \times 10^{-5}$	.04659	.92	$.668 \times 10^{-4}$	$.681 \times 10^{-4}$

Here

$$\sigma_e^2 = \delta_0 / (1 - \delta_1 - \beta_1), \qquad \sigma_x^2 = \sigma_e^2 / (1 - \phi_1^2).$$

and he claims the significance of estimates and observes;

(a) comared to ①, the estimates of ϕ_0 and ϕ_1 do not change,

(b) the estimate of δ_0 in ② is very small and hence the condi-
     tional variance is changing,

(c) σ_e^2 and σ_x^2 are almost same between ① and ②, and

(d) the value of β_1 is close to 1 and the value of $\delta_1 + \beta_1$ is
     very close to 1, but an application of the Dickey=Fuller unit
     root test rejects the hypothesis $\delta_1 + \beta_1 \geq 1$.

The fact (d) will imply the stationarity of the conditional
variance h_t from $\delta_1 + \beta_1 < 1$ and the predictability of variances
(volatilities) from the closeness of $\delta_1 + \beta_1$ to 1. He in fact
explored for the predictability and compared simple forecasts, the
forecasts by an exponentially smoothing model, the forecasts by the
ARCH model and the forecasts by the GARCH model, and observed the
forecasts by the GARCH model is best among them. It is noted that
in these models the conditional mean g_t of x_t is fixed as AR(1)
model as specified above. Also note that in the Dickey=Fuller test
the normality of the series is rather essential.

C. Interest rates

Baba (1990) analyzed monthly data of twenty year Treasury Bond
rates and one month Treasury Bill rates (Solomon Brother's Analytical Rocord of Yield and Yield Spread). As in Campbell and Shiller
(1984), let $h y_t$ be one month holding period yields on the long
term bonds and let r_t be the one month Treasury bill rate. The
excess return process $\{x_t\}$ is analyzed for 1970.1~1985.11 (191
data), which provides the summary statistics;

x	s	Max	Min	b	k
-.06448	3.37886	11.487	-11.926	-.209	4.56

The autocorrelation of $\{x_t\}$ and $\{x_t^2\}$ are as follows.

	r_1	r_2	r_3	r_4	r_5	r_6	r_7	r_8
$\{x_t\}$	.133	-.014	-.174	.010	.095	.095	-.042	-.058
$\{x_t^2\}$	.096	.125	.230	.105	-.024	.222	.018	-.026

Some similar features as we have observed are also observed here (nonnormality, relatively high autocorrelations in $\{x_t^2\}$). The model Baba considered is

$$x_t = \beta_0 + \beta_1 s p_t \cdot \sqrt{h_t} + \varepsilon_t \qquad \varepsilon_t \sim N(0, h_t),$$
$$h_t = \alpha_0 + \alpha_1 \Sigma_{j=1}^{k} \omega_j \varepsilon_{t-j}^2 \qquad \omega_j = (k-j+1)/k(k+1)$$

where $s p_t$ is a yield spread and h_t is conditional variance given Ψ_{t-1}. This model corresponds to the ARCH-M model in (3), and is similar to Campbell and Shiller (1984). The estimates are given as follows.

	α_0	α_1	β_0	β_1
	5.051	0.5371	-0.8034	1.610
t-value	5.2	3.4	-2.6	3.7

For a diagnostic checking for the model the autocorrelations and Box-Pierce tests based on $\hat{u}_t \equiv \hat{\varepsilon}_t / h_t^{1/2}$ are computed and the white noise-ness is supported. But the adequancy of the model was checked in graphs. A remark is that the normality of $\{\hat{u}_t\}$ maybe need to be checked.

5 Some other nonlinear models

Univariate nonlinear models will be useful not only as it stands for univariate series, but also in modelling factors or variance components in a multifactor model. Recently various nonlinear models are proposed and applied in real world, in particular in natural sciences. In general, as is discussed in Section 1, a non-

linear model can be specified in an arbitrary manner, and hence it
will be required that a model should have good performance for a
specific phenomenon and a kind of interpretability.   In this
section we briefly review the following models;
   ① bilinear model,
   ② threshold autoregressive model,
   ③ smoothing transition model.
In addition, in later chapters, various versions of
   ④ random coefficient model
are treated.

Bilinear model

   Good reference books on bilinear models are Granger and Anderson
(1978) and Subba Rao and Gahr (1980).   A general bilinear time
series model is expressed as

$$(5.1) \quad x_t = \sum_{i=1}^{a} \phi_i \, x_{t-i} + \sum_{j=0}^{b} \theta_j \varepsilon_{t-j}$$
$$+ \sum_{k=1}^{c} \sum_{m=1}^{d} \beta_{km} \, x_{t-k} \varepsilon_{t-m},$$

where $\{\varepsilon_t\}$ is iid.   The bilinear model of the form (5.1) is often
expressed as $BI(a, b, c, d)$.   A special case sometimes considered in
applications is a $BI(a, 0, c, 1)$;

$$(5.2) \quad x_t = \sum_{i=1}^{a} \phi_i \, x_{t-i} + \varepsilon_t + \sum_{k=1}^{c} \beta_{k1} \, x_{t-k} \varepsilon_{t-1}$$

(see Subba Rao and Gahr (1980) for an analysis of earthquakes).
A necessary and sufficient condition for the simplest bilinear model

$$(5.3) \quad x_t = \phi_1 x_{t-1} + \varepsilon_t + \beta_{11} \varepsilon_{t-1} x_{t-1},$$

to be stationary is given by Pham and Tran (1981) as

$$\phi_1^2 + \sigma^2 \beta_{11}^2 < \infty \quad \text{with} \quad E(\varepsilon_t^2) = \sigma^2.$$

A sufficient condition for the model (5.2) to be stationary is given by Bhaskara Rao et al (1983) and a further extension is made by Liu and Brockwell (1988). But we have to wait for a report on a successful application of the model to financial time series. The estimation of the model can be made by the ML method or the following two stage least sqaure procedure.

i) Regress x_t on $x_{t-1}, \cdots, x_{t-a}$ with a large and get the coefficients for ϕ_i's in (5.2), and the residuals $\{e_t\}$.

ii) Regress e_t on $x_{t-1} e_{t-1}, \cdots, x_{t-a} e_{t-1}$ for (5.2), and get the coefficients for β_{k1}'s.

Threshold autoregressive model

A threshold autoregressive model with two regimes is given by

$$(5.4) \quad \begin{array}{l} x_t = \phi_{0(1)} + \phi_{1(1)} x_{t-1} + \cdots + \phi_{a(1)} x_{t-a} + \varepsilon_{t(1)} \\ \qquad\qquad\qquad\qquad\qquad\qquad\qquad\quad \text{if} \quad x_{t-d} \leqq r, \\ x_t = \phi_{0(a)} + \phi_{1(2)} x_{t-1} + \cdots + \phi_{a(2)} x_{t-a} + \varepsilon_{t(2)} \\ \qquad\qquad\qquad\qquad\qquad\qquad\qquad\quad \text{if} \quad x_{t-d} > r, \end{array}$$

where x_{t-d}, d and r are respectively called threshold variable, delayed parameter and threshold value. This nonlinear model is proposed by Tong (1978, 1983) and Tong and Lim (1980) as an alternative model for describing periodic time series, limit cycles, amplitude dependent frequencies, jump phenomena, etc. The model may be viewed as an extension of switching regression scheme to time series and it is in fact expressed as

$$(5.5) \quad x_t = \phi_{0(1)} + \Sigma_{j=1}^{a} \phi_{j(1)} x_{t-j} + \varepsilon_{t(1)} \\ \qquad\qquad + [\beta_0 + \Sigma_{j=1}^{a} \beta_j x_{t-j} + \xi_t] \chi_{(r,\infty)}(x_{t-d})$$

where $\beta_j = \phi_{j(2)} - \phi_{j(1)}$ ($j = 0, 1, \cdots, a$), $\xi_t = \varepsilon_{t(2)} - \varepsilon_{t(1)}$ and $\chi_{(r,\infty)}(x_{t-d})$ is 1 if $x_{t-d} > r$ and 0 otherwise. A model selection procedure for this model is given in Tsay (1989) where d is selected based on a nonlinearity test for the TAR model. In the simplest model

$$X_t = \begin{cases} \phi_{1\,(1)}\, X_{t-1} + \varepsilon_{1\,(t)} & \text{if } X_{t-1} \leqq 0, \\ \phi_{1\,(2)}\, X_{t-1} + \varepsilon_{2\,(t)} & \text{if } X_{t-1} > 0, \end{cases}$$

Petruccalli and Woolford (1984) showed that

$$\phi_{1\,(1)} < 1, \quad \phi_{1\,(2)} < 1 \quad \text{and} \quad \phi_{1\,(1)}\, \phi_{2\,(1)} < 1$$

is a necessary and sufficient condition for stationarity.

Smoothing transition autoregressive model

A natural extension of the model (5.5) is clearly obtained by replacing the indicator function χ by a smoothing function F as

$$X_t = \alpha_0 + \Sigma_{j=1}^{a} \alpha_j X_{t-j} + \varepsilon_t \\ + [\beta_0 + \Sigma_{j=1}^{a} \beta_j X_{t-j} + \xi_t] F(X_{t-d})$$

where $F(z)$ is a distribution function on R. One may choose $F(z) = \Phi(z/\gamma)$ where Φ is the standard normal distribution function.

Example: Nikkei 225

Let us model the absolute return process $\{|X_t|\}$ of monthly Nikkei 225 Index as it exhibits a slowly declining and positive autocorrelation structure as we observed in Chapter 2. The slow convergence of autocorrelations $\rho_k = \rho_k(|x|)$ to zero may be related to the time series structure of long-term dependency, which we will discuss in Chapter 13 associated with options. It is noted that $y_t \equiv |x_t|$ may be regarded as $|x_t - \bar{x}|$ as $\bar{x}$ is usually small. Then y_t may be regarded as an estimate of the standard deviation σ_t of x_t at t. Hence modelling $\{y_t\}$ and predicting future values of y's can be associated with the problem of predicting volatilities (see Section 5 of Chapter 12). The data period is from 1965.2 to 1989.12 (monthly). Optimal AR and TAR models are first estimated for 1965.2～1984.8 and 64 predicted

values from the models are obtained. Then one data is added to the
period (i.e. 1965.2~1984.9) to estimate the models and predict 63
future values. Continuing the procedure, we obtain 64 predicted
values for one month ahead prediction, 63 for two months ahead prediction, etc. The autocorrelations (AC) and partial autocorrelations (PAC) of $\{y_t\}$ for the period 1965.2~1984.8 are given as
follows.

	1	2	3	4	5	6	7	8	9	10	11	12
A C	.23	.25	.22	08	.19	.12	.13	.11	.12	.05	.12	.00
PAC	.23	.20	.14	-.04	.13	.03	.06	.01	.05	-.05	.08	-.08

This table shows that the autocorrelations of $\{y_t\}$ are quite big
compared to those of $\{x_t\}$ itself though it is a monthly series.
A procedure of selecting an ARMA model for the period gives us the
following AR(5);

$$y_t = 0.02 + 0.03\, y_{t-1} + 0.12\, y_{t-2} + 0.16\, y_{t-3} + 0.14\, y_{t-5} + \varepsilon_t,$$
$$\sigma = 0.00078.$$

While, the TAR model selection procedure in the University of
Chicago software gives

$$y_t = \begin{cases} a_0 + a_1\, y_{t-1} + a_2\, y_{t-2} + a_3\, y_{t-3} + \varepsilon_{1t} & \text{if } y_{t-12} \leq 0.034, \\ b_0 + b_1\, y_{t-1} + b_2\, y_{t-2} + b_3\, y_{t-3} + \varepsilon_{2t} & \text{if } y_{t-12} > 0.034, \end{cases}$$

where estimated parameters are given in the table.

i	0	1	2	3	σ	N
a_i	.020	.10	.18	.14	.000704	168
b_i	.030	-.03	.13	.16	.000826	119

The two models are not quite different in error standard deviations.

To see a difference of performances, the prediction MSE (mean squares errors)'s are computed for k month ahead values ($k=1,2,$ $\cdots,10$). The result is given in the next table.

k	1	2	3	4	5	6	7	8	9	10
AR(5)	852	844	803	857	849	882	878	890	913	916
TAR	810	811	818	875	857	874	889	894	924	915

The numbers are MSE$\times 10^6$.

In each MSE, $65-k$ predicted values are compared to realized values. From this table, it seems difficult to conclude that the TAR is better than the AR model though the TAR is 5% better in the MSE for one month ahead prediction.

Exercises

1. State your opinion on the role of the assumption of stationarity for an economic model and compare it with the argument in Section 1.

2. In (2.3), show that $\sum_{j=0}^{k} \varepsilon_{t-j}$ diverges as $k \to \infty$.

3. In Taylor model (3.2) with (3.3), show that $u_t v_t = u_t^* v_t^*$ implies $u_t = u_t^*$ and $v_t = v_t^*$ (a.e.).

4. Suppose a process $\{v_t\}$ with $v_t > 0$ follows a log AR(1) model; $\log v_t - \alpha = \phi(\log v_{t-1} - \alpha) + \eta_t$ with ϕ and $\{\eta_t\} \sim$ iid $N(0, \beta^2(1-\phi^2))$.
   Show that (i) $\log v_t \sim N(\alpha, \beta^2)$, (ii) $E(v_t^r) = \exp(r\alpha + \frac{1}{2}r^2\beta^2$
   (iii) $\rho_{\tau, v} = [\exp(\beta^2 \rho_{\tau, \log v}) - 1]/[\exp(\beta^2) - 1]$.

5. Taylor (1986) lists a computer program for his model. If it is available, analyze a financial time series in a line with the argument in Section 3 and test on the RWH.

6. In ARCH model (4.1), assume $g_t \equiv 0$ for all t and $h_t = \delta_0 + \delta_1 \varepsilon_{t-1}^2$. Let $z_t = h_t^{1/2} \varepsilon_t$ with $\sigma_\varepsilon^2 = \text{Var}(\varepsilon_t)$. Show that
   (i) the process $\{z_t\}$ is stationary with finite variance if and

only if $\sigma_\varepsilon^2 \delta_1 < 1$, (ii) the unconditional variance is then $\sigma_\varepsilon^2 \delta / (1 - \sigma_\varepsilon^2 \delta_1)$ and (iii) the fourth moment is finite if $3\sigma_\varepsilon^4 \delta_1^2 < 1$.

7. Study on the proof of Lemma 2.1.

CHAPTER 4

MULTIVARIATE FINANCIAL TIME SERIES MODELS

1 Multifactor models

Most popular multivariate models describing price variations of financial assets in practice are so-called multifactor models. The popularity will be not only because multifactor models are compatible with the framework of finance theories to some extent but also because the models are easy to understand and treat statistically. In fact, the models are of great applicability and of many varieties. In this chapter, we overview some formulations of multifactor models.

Suppose financial assets consist of p stocks and let x_{it} be the return of the i-th stock at time t. Then finance theories often treat a multifactor model of the form

$$(1.1) \quad x_{it} = \mu_{it} + \alpha_{i1} f_{1t} + \cdots + \alpha_{iq} f_{qt} + \varepsilon_{it} \quad (i=1, \cdots, p),$$

where f_{jt}'s ($j=1, \cdots, q$) are q factors affecting commonly the p returns x_{it}'s ($i=1, \cdots, p$). For example, in the CAPM (capital asset pricing model), the model (1.1) with $q=1$ is derived as a market equilibrium model under the assumptions of homogeneous expectations, perfect information, etc. (see Chapter 7). On the other hand, in the APT (arbitrage pricing theory) the model (1.1) is assumed as a variation model for returns and an expression of the expected return $E(x_{it})$ is derived with concepts of arbitration and of diversified portfolios. These theories are static and use one-period models, and hence do not argue about the time series structure of the common variation components (factors) f_{jt}'s. In fact, it is assumed in a one-period model;

a) people at time 0 have perfect information and form the same expectations on the distribution of ($x_{11}, \cdots, x_{p1}$) at time 1. Hence the means and covariances of the p returns x_{i1}'s are known,

58

b) at time 1 all the investment activities are closed.

However, as has been discussed in Chaper 1, our investment time horizons in practice are quite different from people (fund) to people (fund) and hence we need to forecast multiperiod returns over some future periods, in which case the time series structure of factors f_{jt}'s is very important. Though a multiperiod (intertemporal) CAPM is also developed, it is often assumed that the process of

$$(1.2) \quad x_t = (x_{1t}, \cdots, x_{pt})' : p \times 1$$

is an iid process. But as has been observed, x_{it}'s are not independent. Further, investors change their investment behaviors as prices change and their investment decisions in reality depend on the information of price levels as well as on the information of returns. Also it will be difficult to develop an APT with time series structures for factors f_{jt}'s even if the model (1.1) is accepted.

Taking it for granted that an investment is a commitment to future, what is important in using data is to extract the information contained in data which is relevant to forecasting the future movement of prices. There the forms of data or variates are not predetermined. In fact, investors or analysts use different investment measures and various technical methods, which are often nonlinear in price levels and in returns, and they forecast price levels through the relationships between prices and the measures they rely on. In addition, price levels are important notions not only for investors but also for fundamentals analysts. Hence in our argument below, x_{it}'s can be any forms of given variates, and the variates may be any variables an analyst thinks relevant.

2 A review on some basic multivariate time series models

Let $\{x_t\}$ be a p-dimensional time series where x_t is a p-

dimensional vector (as in (1.2)) of variables of interest. Let μ_t $= E(x_t)$ be the mean vector and let

$$(2.1) \quad \Sigma_t(k) = \text{Cov}(x_t, \ x_{t-k})$$
$$= E(x_t - \mu_t)(x_{t-k} - \mu_{t-k})'$$

be the cross-serial covariance matrix of x_t and x_{t-k}. The (i, j) element of $\Sigma_t(k)$ is

$$(2.2) \quad \sigma_{ijt}(k) = E(x_{it} - \mu_{it})(x_{jt-k} - \mu_{jt-k}).$$

If $i = j$, then $\sigma_{iit}(k)$ is the autocovariance of x_{it} and x_{it-k}. In general, $\Sigma_t(k)$ depends on t. But if μ_t and $\Sigma_t(k)$ (or equivalently $\sigma_{ijt}(k)$) is independent of t, then we call $\{x_t\}$ a stationary process and write $\mu_t \equiv \mu$ and $\Sigma_t(k) \equiv \Sigma(k)$.

Suppose $\{x_t\}$ is stationary. Then it is easy to see $\sigma_{ij}(-k) = \sigma_{ji}(k)$ and hence

$$(2.3) \quad \Sigma(k)' = \Sigma(-k).$$

Also the lag k cross-autocorrelation matrix of x_t is defined by

$$(2.4) \quad \Gamma(k) = (\rho_{ij}(k)) \qquad \text{with}$$
$$\rho_{ij}(k) = \sigma_{ij}(k)/(\sigma_{ii}(0)\sigma_{jj}(0))^{1/2}.$$

Of course, $\rho_{ij}(k)$ measures the degree of linearity of x_{it} and x_{jt-k} and hence if $\rho_{ij}(k)$ with $k>0$ is close to 1, x_{jt-k} will "causes" x_{it}. Note $\Sigma(0) = \text{Var}(x_t)$. For given data $\{x_1, \cdots, x_T\}$, $\Sigma(k)$ is estimted by

$$(2.5) \quad \hat{\Sigma}(k) = \frac{1}{T}\Sigma_{t=k+1}^{T}(x_t - \bar{x})(x_{t-k} - \bar{x})' \qquad \text{with}$$
$$\bar{x} = \frac{1}{T}\Sigma_{t=1}^{T} x_t.$$

VAR(m) model

A multivariate extension of univariate AR model is a VAR

(vector autoregressive) model of the form

$$(2.6) \quad z_t = \Phi_1 z_{t-1} + \Phi_2 z_{t-2} + \cdots + \Phi_m z_{t-m} + \varepsilon_t,$$

where $z_t = x_t - \mu_t$, Φ_i's are $p \times p$ coefficient matrices and ε_t is an error term. The process of $\{\varepsilon_t\}$ is assumed to be a white noise;

$$(2.7) \quad E(\varepsilon_t) = 0, \quad \mathrm{Var}(\varepsilon_t) = \Omega \quad \text{and}$$
$$\mathrm{Cov}(\varepsilon_t, \varepsilon_{t-k}) = 0 \quad \text{for any } t \text{ and } k \neq 0.$$

The model (2.6) is stationary if all the roots of

$$(2.8) \quad |\omega^m I_p - \omega^{m-1} \Phi_1 - \cdots - \Phi_m| = 0$$

are less than one in absolute values. A VAR(m) model involves $p^2 \times m + \frac{1}{2}p(p+1)$ parameters, and Φ_i's are estimated by the least square method or the maximum likelihood method. Let β_i be a vector consisting of the column vectors of $\hat{\Phi}_i$ and let $\beta = (\beta_1', \cdots, \beta_m')'$. Then the least squares estimator $\hat{\beta}$ of β is asymptotically distributed as

$$(2.9) \quad \sqrt{T}(\hat{\beta} - \beta) \sim N(0, \Sigma(0)^{-1} \otimes \Omega)$$

when $T \to \infty$. Here $\Sigma(0)$ is the covariance matrix of z_t.

VARMA(m, n) model

As a multivariate extension of univariate ARMA (autoregressive moving average) model, a VARMA(m, n) model of order (m, n) is of the form

$$(2.10) \quad z_t = \Phi_1 z_{t-1} + \cdots + \Phi_m z_{t-m} + \varepsilon_t + \Theta_1 \varepsilon_{t-1} + \cdots + \Theta_n \varepsilon_{t-n}$$

where $\{\varepsilon_t\}$ is a white noise satisfying (2.7). A VARAMA(m, n) involves $p^2 \times (m + n) + \frac{1}{2}p(p+1)$ parameters.

In these multivariate models, model identification is not easy

to carry out, which will be discussed later again. A VARMA model
is not usually used because of a computational difficulty in estimation and model identification. In a VAR model, model selection
criteria for identification commonly used are the AIC (Akaike's
information criterion) and the SBIC (Schwarz's Bayesian information
criterion):

$$\text{AIC} = \log|\hat{\Omega}| + 2\,m\,p^2/\,T,$$
(2.11)
$$\text{SBIC} = \log|\hat{\Omega}| + m\,p^2\log T/\,T,$$

where $\hat{\Omega}$ is the sample covariance matrix of the least squares
residuals $\hat{\varepsilon}_t$'s.

A VAR model takes into account all the conceivable linear relations between a single variable z_{it} at time t and the past
variables z_{t-1}, z_{t-2}, $\cdots$, z_{t-m}. However, because of the fact, too
many unknown parameters are introduced into the model, which is as
many as $p^2 \times m$ for coefficients and $p(p+1)/2$ for covariance
matrix Ω, and the model is very likely to be destabilized though
model fit is often very high apparently. This creates the problem
of the lack of degrees of freedom in data and the problem of low
performance of prediction. In fact it is often observed:
1) An addition of a new variable $z_{p+1\,t}$ to a given set of varia-
    bles ($z_{1\,t}, \cdots, z_{p\,t}$) often causes a great change of the results
    via the models. In other words, it is important to choose a
    proper set of variables carefully in order to obtain a stable
    performance.
2) A change of transformation of a variable in ($z_{1\,t}, \cdots, z_{p\,t}$) also
    sometimes affects the total result.
3) Because parameters increase by p^2 as lags increase by one,
    model performance greatly changes with lag m. The model
    selection criteria as stated above are not quite effective in
    multivariate cases because the number of parameters increases
    not one by one, but p^2 by p^2. In other words, there may be

many insignificant parameters involved in each coefficient matrix Φ_i, but they are hard to identify in the process of model identification. When p is large, the identification of zero parameters in each Φ_i are almost impossible.

Tiao and Tsay (1989) have made a great deal of contributions to solving the two major difficulties in applying a VARMA model;

(1) overflow of parameters the estimates of which are highly correlated and

(2) lack of identifiable models.

The first problem has been discussed. To illustrate the second problem, suppose that $\{x_t\}$ is generated by the VARMA(1,1) model;

$$(2.12) \quad x_t = \Phi_1 x_{t-1} + \varepsilon_t + \Theta_1 \varepsilon_{t-1} \quad \text{with}$$

$$\Phi_1 = \begin{pmatrix} 3 & -1 \\ 6 & -2 \end{pmatrix} \quad \text{and} \quad \Theta_1 = \begin{pmatrix} 0.5 & -0.5 \\ 1 & -1 \end{pmatrix}.$$

In this expression, there are 8 nonzero coefficients. But it is easy to see that the ranks of Φ_1 and Θ_1 are 1 and that transforming x_t into $z_t = T x_t$ with

$$(2.13) \quad T = \begin{pmatrix} 2 & -1 \\ -1 & 1 \end{pmatrix}$$

yields

$$(2.14) \quad z_t = \Phi_1^* z_{t-1} + \xi_t + \Theta_1^* \varepsilon_{t-1} \quad \text{with} \quad \xi_t = T\varepsilon_t,$$

$$\Phi_1^* = T\Phi_1 T^{-1} = \begin{pmatrix} 0 & 0 \\ 2 & 1 \end{pmatrix} \quad \text{and} \quad \Theta_1^* = T\Theta_1 T^{-1} = \begin{pmatrix} 0 & 0 \\ 0 & 0.5 \end{pmatrix}$$

In (2.14) there are only three nonzero coefficients, and $z_{1t} = \xi_{1t}$, which is ARMA(0,0), i.e., white noise, and

$$z_{2t} = 2 z_{1t-1} + z_{2t-1} + \xi_{2t} + 0.5\xi_{2t-1}.$$

In other words, the model (2.12) admits a simpler expression by a suitable transformation T and in particular $z_{1t} = (2,-1)x_t$ via T in (2.13) contains no nonzero coefficients, which is in contrast to the fact that x_{1t} contains 4 nonzero coefficients in (2.12). In general, for a given VARMA(n, m) model (2.10) $y_t = v_0' \, x_t$ is said to follow a scaler component model (SCM) of order (n_1, m_1) if $0 \leqq n_1 \leqq n, \ 0 \leqq m_1 \leqq m$,

$$
\begin{array}{llll}
(2.15) & v_0' \, \Phi_{n_1} \neq 0, & v_0' \, \Phi_k = 0 & (k = n_1 + 1, \cdots, n) \quad \text{and} \\
 & v_0' \, \Theta_{m_1} \neq 0, & v_0' \, \Theta_k = 0 & (k = m_1 + 1, \cdots, m).
\end{array}
$$

In fact, Tiao and Tsay (1989) gave a more general definition and obtained a statistical procedure for finding an SCM structure.

Furthermore, it is also possible to rewrite the reduced model (2. 14) as

$$
(2.16) \quad z_t = \Phi_1^{**} z_{t-1} + \xi_t + \Theta_1^{**} \xi_{t-1} \quad \text{with}
$$

$$
\Phi_1^{**} = \begin{pmatrix} 0 & 0 \\ 0 & 1 \end{pmatrix} \quad \text{and} \quad \Theta_1^{**} = \begin{pmatrix} 0 & 0 \\ 2 & -0.5 \end{pmatrix}
$$

because $z_{1t} = \xi_{1t}$. Therefore the model (2.14) is not uniquely identifiable. Tiao and Tsay (1989) call the two VARMA(1.1) models in (2.14) and (2.16) exchangeable in the sense that they have different coefficients but yields the same probabilistic structure of z_t. Consequently it is necessary to identify classes of exchangeable models in finding an SCM structure. An SCM structure (2.15) is reduced to the relation

$$
(2.17) \quad \Sigma(k) v_0 + \Sigma(k-1) v_1 + \cdots + \Sigma(k - n_1) v_{n_1} = 0
$$
$$
\text{for } k > m_1
$$

for some $v_1, \cdots, v_{n_1}$ with $v_{n_1} \neq 0$. Hence a main procedure for finding an SCM structure is to check whether the autocovariance matrices $\{\hat{\Sigma}(k)\}$ satisfy (2.17), where $\hat{\Sigma}(k)$ is given by (2.5). They use an interesting argument for the procedure by using canonical correlation analysis. But a difficulty in applications to p-dimensional financial time series with p being not small is that T cannot be very large, which we have discussed in Chapter 3 and they also pointed out. In other words, $\hat{\Sigma}(k)$'s will not be very reliable to identify an SCM structure. A further study including empirical results will be expected.

3 State space approach and Kalman Filter model

Most multivariate time series models can be put in the framework of a state space model which consists of a state space equation and an observation equation. For example, for the multifactor model (1. 1) with factors f_{jt}'s following time series processes, the process of

$$(3.1) \quad f_t = (f_{1t}, \cdots, f_{qt})' : \quad q \times 1$$

is formulated as a state space equation and the f_{jt}'s are regarded as the state variables which can be unobservable. While, the model (1.1) which relates the state variables to the observable variables x_{it}'s is called an observation equation. This viewpoint is often taken in engineering. In general the two equations can be expressed as follows.

I **State Space Equation:** an equation which describes the time
   series structure of a state variable vector f_t;

$$(3.2) \quad f_t = F(f_{t-1},\ f_{t-2},\ \cdots,\ f_{t-k} : \eta_t),$$
$$\{ \eta_t \}: \quad q \text{ dimensional white noise.}$$

II **Observation Equation:** an equation which relates f_t to x_t;

$$(3.3) \quad x_t = G(f_t,\ f_{t-1},\ \cdots,\ f_{t-r} : \varepsilon_t),$$
$$\{ \varepsilon_t \}: \quad p \text{ dimensional white noise.}$$

In applications, it is often the case in a state space approach that
  1) these equations are linear,
  2) state vector f_t is unobservable,
  3) the information concerning f_t is obtained only through x_t,

and based on the observations $\{ x_s : s = 1, \cdots, t \}$ up to t, the equations are estimated and the movement of state vector f_t is aimed to capture and forecast. In a typical engineering situation, f_t's are the target variables to identify and model. But they are not directly observable without noises (observational errors), and their movements are observalbe only through the observation equation (3.3) which transforms the information on state vectors f_{t-j}'s into an observation x_t with noise ε_t. In our applications the movements of x_t's are of greater interest as in (1.1). In the engineering terminology, the estimation of past states $f_{t-\ell}$ ($\ell = 1, 2, \cdots$) is called **smoothing**, the estimation of present state f_t is called **filtering**, and the estimation of future states $f_{t+\ell}$ ($\ell = 1, 2, \cdots$) is of course **prediction**, where data is $\{ x_s : s = 1, \cdots, t \}$. Though there are many variations of the model, a typical linear state space model is of the form

$$
\begin{aligned}
(3.4) \quad f_t &= B_{1t} f_{t-1} + \cdots + B_{kt} f_{t-k} + \eta_t & (B_{it} : q \times q), \\
x_t &= A_{0t} f_t + A_{1t} f_{t-1} + \cdots + A_{rt} f_{t-r} + \varepsilon_t & (A_{it} : p \times q),
\end{aligned}
$$

where x_t in (3.4) is regarded as $x_t - \mu_t$ with $\mu_t = E(x_t)$. In other words, state vectors f_t's follow a VAR(k) model and observation vector x_t depends on present and past state vectors.

The model (3.4) is expressed as

$$
\begin{aligned}
(3.5) \quad f_t &= B_t f_{t-1} + \eta_t, \\
x_t &= A_t f_t + \varepsilon_t
\end{aligned}
$$

by redefining the variables f_t and x_t, which will be soon demonstrated in examples.

Example 1 Let an AR(q) model be

$$
z_t = \phi_1 z_{t-1} + \cdots + \phi_q z_{t-q} + u_t.
$$

Define

$$f_t = \begin{pmatrix} z_t \\ z_{t-1} \\ \cdot \\ \cdot \\ \cdot \\ \cdot \\ z_{t-q+1} \end{pmatrix} : q \times 1, \quad B = \begin{pmatrix} \phi_1, & \phi_2, & & \phi_q \\ 1 & 0 & \cdots\cdots & 0 \\ 0 & & & \\ \cdot & & & \\ \cdot & & & \\ \cdot & & & 0 \\ 0 & \cdots\cdots\cdots & 0 & 1 & 0 \end{pmatrix} : q \times q,$$

$$\eta_t = \begin{pmatrix} u_t \\ 0 \\ \cdot \\ \cdot \\ 0 \end{pmatrix} : q \times 1, \quad x_t = f_{1t} = z_t, \quad A = [1, 0, \cdots, 0] : 1 \times q.$$

Then we obtain

$$f_t = B f_{t-1} + \eta_t \quad \text{and} \quad x_t = A f_t,$$

which is a special case of (3.5). An ARMA model is similarly expressed as a state space model. It is noted that the expression is not unique.

Example 2 Time-Varying Coefficient Model

Suppose the return r_{it} of the i-th stock at time t follows the following model

(a) $r_{it} = \beta_{it} r_{mt} + u_{it}$,

(b) $\beta_{it} = \phi_{i1} \beta_{it-1} + \cdots + \phi_{ik} \beta_{it-k} + v_{it}$,

where r_{mt} is the return of a stock index at time t. The model (a) is usually called a market model and discussed in details in Chapter 7. Usually a market model has a constant coefficient, but here the coefficient β_{it} is assumed to follow an AR(k) model as in (b). Let

$$\beta_{it} = (\beta_{it}, \cdots, \beta_{it-k+1})' : k \times 1,$$
$$a_t = (r_{mt}, 0, \cdots, 0) : 1 \times k,$$

$$
B_i = \begin{pmatrix} \phi_{i1}, & \cdots\cdots, & \phi_{ik} \\ 1 & 0 & \\ 0 & & \\ \cdot & & \\ \cdot & & \\ \cdot & & \\ 0 & \cdots\ 0\ \ 1 & 0 \end{pmatrix}, \qquad \eta_{it} = \begin{pmatrix} v_{it} \\ 0 \\ \cdot \\ \cdot \\ \cdot \\ 0 \end{pmatrix}
$$

and $\varepsilon_{it} = u_{it}$. Then it follows that

(b)': $\quad \beta_{it} = B_i \beta_{it-1} + \eta_{it}$,

(a)': $\quad r_{it} = a_t \beta_{it} + \varepsilon_{it}$.

This model is of the form (3.5) with $f_t = \beta_{it}$ and $x_t = r_{it}$. However, in this model a_t for A depends on time. This does not create a problem because a_t is an observable vector and contains no unknown parameters.

For N stocks, define

$$
\beta_t = \begin{pmatrix} \beta_{1t} \\ \cdot \\ \cdot \\ \cdot \\ \beta_{Nt} \end{pmatrix}, \quad \eta_t = \begin{pmatrix} \eta_{1t} \\ \cdot \\ \cdot \\ \cdot \\ \eta_{Nt} \end{pmatrix}, \quad B = \begin{pmatrix} B_1 & & 0 \\ & \cdot & \\ & & \cdot \\ 0 & & B_N \end{pmatrix},
$$

$$
A_t = \begin{pmatrix} a_t & 0 & \cdots & 0 \\ 0 & a_t & \cdots & 0 \\ \cdot & & & \\ \cdot & & & \\ 0 & \cdots\cdots & 0 & a_t \end{pmatrix} = I \otimes a_t, \quad r_t = \begin{pmatrix} r_{1t} \\ \cdot \\ \cdot \\ \cdot \\ r_{Nt} \end{pmatrix}, \quad \varepsilon_t = \begin{pmatrix} \varepsilon_{1t} \\ \cdot \\ \cdot \\ \cdot \\ \varepsilon_{Nt} \end{pmatrix}.
$$

Then a state space model of N stocks is expressed as

(b)": $\quad \beta_t = B \beta_{t-1} + \eta_t$,

(a)": $\quad r_t = A_t \beta_t + \varepsilon_t$,

which is clearly of the form (3.6).

Kalman algorithm

For given past and present data $\{ x_1, \cdots, x_t \}$, we need to obtain a *smoother* estimating the past states $f_{t-\ell}$ ($\ell = 1, 2, \cdots$), a *filter* estimating the present state f_t and a *predictor* estimating the future states $f_{t+\ell}$ ($\ell=1,2,\cdots$).   Kalman gave a recurrsive formula under the following assumption.

Assumption of Kalman Filtering

(i) $\quad f_t = B_t f_{t-1} + \eta_t,\qquad \eta_t$ iid $N(0,\ \Omega)$,

$\quad\quad x_t = A_t f_t + \varepsilon_t,\qquad \varepsilon_t$ iid $N(0,\ \Phi)$,

where $\{\eta_t\}$ and $\{\varepsilon_t\}$ are independent, and

(ii) $\quad A_t,\quad B_t\ \ (t=0,1,2,\cdots),\quad \Omega$, and Φ are known.

To give his formula, let

$$(3.6)\quad P(\widetilde{f}_{t|t}) = E[(f_t - \widetilde{f}_{t|t})(f_t - \widetilde{f}_{t|t})' \mid X_t]: q\times q\ \ \text{with}$$

$\widetilde{f}_{t|s}$: an estimator of f_t based on $\{x_1,\cdots,\ x_s\}$,

which is the conditional mean squared matrix of $\widetilde{f}_{t|t}$ for f_t given $X_t=(x_1',\ x_2',\ \cdots,\ x_t')'$.   Let $\widehat{f}_{t|t}$ be the best filter minimizing $P(\widetilde{f}_{t|t})$ and let

$$(3.7)\quad P_{t|t} \equiv P(\widehat{f}_{t|t}) = \min_{\widetilde{f}_{t|t}} P(\widetilde{f}_{t|t}).$$

Theorem 3.1. (Kalman) The best predictor of f_{t+1} at time t is

$$(1)\qquad \widehat{f}_{t+1|t} = B_{t+1}\,\widehat{f}_{t|t},$$

where $\widehat{f}_{t|t}$ is the best filter of f_t at time t and given by

$$(2)\qquad \widehat{f}_{t|t} = \widehat{f}_{t|t-1} + P_{t|t-1}\,A_t'\,F_t^{-1}[x_t - A_t\,\widehat{f}_{t|t-1}]$$

with

$$(3)\qquad F_t = A_t\,P_{t|t-1}\,A_t' + \Phi,$$

$$(4)\qquad P_{t|t-1} = B_t\,P_{t-1|t-1}\,B_t' + \Omega\ \ \text{and}$$

$$(5)\qquad P_{t|t} = P_{t|t-1} - P_{t|t-1}\,A_t'\,F_t^{-1}\,A_t\,P_{t|t-1}.$$

The proof of this theorem is given in Appendix.   This theorem provides the following computational algorithm for each additional data.

i) Choose initial estimates $\hat{f}_{0|0}$ for f_0 and $P_{0|0} = P(\hat{f}_{0|0})$
   (say, $\hat{f}_{0|0} = 0$ and $P_{0|0} = I$) and obtain $\hat{f}_{1|0}$ from (1).

ii) Compute $P_{1|0}$ from (4), F_1 from (3) and then the best filter
   $\hat{f}_{1|1}$ from (2) with x_1 whose mean squared error $P_{1|1}$ is
   obtained from (5).

iii) Obtain the best predictor $\hat{f}_{2|1}$ at time 1 from (1).

iv) Compute $P_{2|1}$ from (4), F_2 from (3) and $\hat{f}_{2|2}$ with x_2. Also
   obtain $P_{2|2}$ from (5).

Repeating this procedure, for given $\{ x_1, \cdots, x_t \}$ we can obtain the
best filter $\hat{f}_{t|t}$ of f_t and the best predictor $\hat{f}_{t+1|t}$ of f_{t+1} at
time t. The best predictor $\hat{f}_{t+k|t}$ of k period ahead state f_{t+k}
is given by

$$\hat{f}_{t+k|t} = B_{t+k} B_{t+k-1} \cdots B_{t+1} \hat{f}_{t|t}.$$

Therefore if we know the coefficient matrices A_t and B_t and
covariance matrices Ω and Φ, then the state variables f_t's are
estimated by the above algorithm. However, in general they are
unknown and need to be estimated also on the basis of the same data.
The maximum likelihood method is used in the estimation. Let the
prediction error by predicting x_t by $A_t \hat{f}_{t|t-1}$ be

(3.8) $\nu_t = x_t - A_t \hat{f}_{t|t-1}.$

Then the log-likelihood function based on $\{ x_1, \cdots, x_T \}$ is given by

$$\log L = -\frac{T}{2} \log 2\pi - \frac{1}{2} \Sigma_{t=1}^{T} \log | F_t | - \frac{1}{2\sigma^2} \Sigma_{t=1}^{T} \nu_t' F_t^{-1} \nu_t.$$

This is a nonlinear function of unknown parameters and hence maximi-
zed by a nonlinear optimization technique.

A.   Regression model and Cusum test

A usual regression model given by

$$y_t = z_t' \gamma + \varepsilon_t \quad \text{with}$$

$$E(\varepsilon_t) = 0 \quad \text{and} \quad \text{Cov}(\varepsilon_t, \ \varepsilon_s) = \sigma^2 \delta_{ts} \quad (\delta_{tt} = 1, \ \delta_{ts} = 0$$
$$(t \neq s))$$

is also regarded as a state space model with

$$\boldsymbol{\gamma}_t = \boldsymbol{\gamma}_{t-1} \quad (\equiv \boldsymbol{\gamma}) \quad \text{and} \quad \eta_t \equiv 0,$$

where $\boldsymbol{z}_t = (z_{1t}, \cdots, z_{pt})'$ and $\boldsymbol{\gamma} = (\gamma_1, \cdots, \gamma_p)'$. Here $f_t \equiv \boldsymbol{\gamma}$, $B_t = I$, $\Omega = 0$, $x_t = y_t$, $A_t = z_t'$, $\varepsilon_t = \varepsilon_t$ and $\Phi = \sigma^2$ in (3.7). Hence the Kalman algorithm in the theorem becomes

$$\hat{\boldsymbol{\gamma}}_{t|t} = \hat{\boldsymbol{\gamma}}_{t|t-1} + P_{t|t-1} \, \boldsymbol{z}_t \, F_t^{-1} (y_t - \boldsymbol{z}_t' \, \hat{\boldsymbol{\gamma}}_{t|t-1}),$$
$$P_{t|t-1} = P_{t-1|t-1},$$
$$P_{t|t} = P_{t|t-1} - P_{t|t-1} \, \boldsymbol{z}_t \, F_t^{-1} \, \boldsymbol{z}_t' \, P_{t|t-1}, \quad \text{and}$$
$$F_t = \boldsymbol{z}_t' \, P_{t|t-1} \, \boldsymbol{z}_t + 1.$$

Here for initial values take $\hat{\boldsymbol{\gamma}}_{p|p}$ to be the OLSE (ordinary least squares estimator) based on $(y_t, \ z_t)$'s $(t-1, \cdots, p)$ and $P_{p|p}$ to be $(Z_p' Z_p)^{-1}$ with $Z_s = (z_1, \cdots, z_s)'$. Then the above algorithm gives the OLSE $\hat{\boldsymbol{\gamma}}_{t|t}$ based on $(y_s, \ z_s)$'s $(s=1, \cdots, t)$. The predictive residual is

$$\hat{\varepsilon}_t = y_t - \boldsymbol{z}_t' \, \hat{\boldsymbol{\gamma}}_{t|t-1} \quad \text{with} \quad \hat{\boldsymbol{\gamma}}_{t|t-1} = (Z_{t-1}' \, Z_{t-1})^{-1} \, Z_{t-1}' \, y_{t-1},$$

where $y_t = (y_1, \cdots, y_t)'$. And σ^2 is estimated by

$$\hat{\sigma}^2 = \frac{1}{T-p} \Sigma_{t=1}^{T} \hat{\varepsilon}_t^2 / F_t \quad \text{with} \quad F_t = \boldsymbol{z}_t' (Z_{t-1}' \, Z_{t-1})^{-1} \, \boldsymbol{z}_t + 1.$$

Define the standardized residuals by

$$e_t = (y_t - \boldsymbol{z}_t' \, \hat{\boldsymbol{\gamma}}_{t|t-1}) / \sqrt{F_t} \qquad t = p+1, \cdots, T.$$

These residuals are often used to construct a test for the hypothesis of the constancy of parameters

$$H: \boldsymbol{\gamma}_t = \boldsymbol{\gamma}_{t-1} \equiv \boldsymbol{\gamma} \quad \text{and} \quad \sigma_t^2 = \sigma_{t-1}^2 \equiv \sigma^2.$$

Brown, Durbin and Evans (1975) proposed the Cusum (cumulative sum) test for $\gamma_t \equiv \gamma$ which rejects for large absolute values of

$$W_t = \Sigma_{s=p+1}^{t} e_s.$$

The asymptotic null distribution of W_t is normal with mean 0,

$$\text{Var}(W_t) = t - p, \quad \text{and} \quad \text{Cov}(W_t, W_s) = \min(t, s) - p.$$

Hence the hypothesis $H: \gamma_t \equiv \gamma$ is rejected if

$$|W_t| > c\sqrt{T\text{-}p} + 2\, c(t - p)/\sqrt{T\text{-}p},$$

where $c=1.143$, 0.984 and 0.850 for significance levels 1%, 5% and 10% respectively. This test is not sensitive to the change in the variances of error terms ε_t's and in its derivation $\sigma_t^2 \equiv \sigma^2$ is assumed.

Another test they proposed is the Cusumsq (cumulative sum of squares) test based on the statistic

$$U_t = \Sigma_{s=p+1}^{t} e_s^2 / \Sigma_{s=p+1}^{T} e_s^2.$$

Under the hypothesis $H: \gamma_t \equiv \gamma$ and $\sigma_t^2 \equiv \sigma^2$, U_t is distributed with mean $(t - p)/(T - p)$ and H is rejected if

$$\left| U_t - \frac{t\text{-}p}{T\text{-}p} \right| > c_0 \qquad \text{for some} \quad t = p+1, \cdots, T.$$

The Ljung-Box test is often used for testing the hypothesis of no autocorrelation of ε_t's, which rejects for large values of

$$Q_m = (T-p)(T-p+2)\Sigma_{k=1}^{m} \hat{\rho}_k^2 / (T-p-k),$$

where $\hat{\rho}_k = \Sigma_{t=p+1+k}^{T} e_{t-k} e_t / \Sigma_{t=p+1}^{T} e_t^2$. The null distribution of Q_m is approximated by χ_m^2.

In applications, these tests are often applied to testing the stability of the CAPM or market model. Knif (1989) applied these tests together with other tests to the Swedish stock returns in the market model and observed the instability of the parameters in most stocks.

B. MMI cash-futures spread on October 19 and 20, 1987

Bassett, France and Pliska (1990) (abbreviated as BFP (1990)) applied the Kalman Filter technique to filtering the minute by minute prices of some untraded stocks in the MMI (Major Market Index) on the 19-th and 20-th of October, 1987, and estimated the value of the Index to evaluate the spread between the MMI and the MMI cash-futures. The MMI consists of 20 stocks;

> Merck & Co., IBM, Phillips Morris, Dow Chemincal, Procter & Gamble, Du Pont, Johnson & Johnson, 3M, GM, Kodak GE, Int. Paper, Exxon, Coca Cola, Chevron, Mobil Sears, USX, AT & T, American Exp.

On the 19-th, there was no trading for most of these stocks for the first two hours after the opening, but thereafter they were continuously traded. On the 20-th, trading was started. But at 10:00 the trading of most stocks was stopped, and at 12:25 trading was retrieved. Hence the MMI could not directly be observed and it needed to be estimated not only for its own purpose but also for the cash-settlement of the MMI futures.

To estimate the Index, BFP (1990) used the Kalman filtering as follows. Let

$$f_t = (f_{1t}, \cdots, f_{20t})' : 20 \times 1 \qquad (0 \leq t \leq 390 \text{ minutes})$$

be the price vector of the 20 stocks, and assume as the process of f_t the random walk model

$$(3.7) \quad f_{t+1} = f_t + \eta_t \text{ with } \eta_t \text{ iid } N(0, \Omega),$$

where f_t is regarded as a state vector. Further let k denote the number of stocks whose prices were observed at t and let $j(1), \cdots, j(k)$ denote the corresponding stock prices, where $0 \leq k \leq 20$.

Then the observation equation is defined by

(3.8) $x_t = B_t f_t + \varepsilon_t$ with ε_t iid $\sim N(0, \Phi)$,

where $x_t = (x_{t(1)}, \cdots, x_{t(k)})': k \times 1$ and

$$B_t = (B_t(i, j)): k \times 20 \quad \text{with} \quad B_t(i, j) = \begin{array}{ll} 1 & j = j(i) \\ 0 & j \neq j(i). \end{array}$$

It is noted that k itself depends on time t, but the recurrsive formula holds even if k depends on t. Also the reason why ε_t appears as an error though x_t in (3.8) is the vector of directly observable prices is that there are bid-ask spreads and tick size errors, as BFP stated. Then assuming the independence of η_t and ε_t, (3.7) and (3.8) form the Kalman filter equations with $A_t = I$, and hence applying Theorem, we can obtain the estimate

(3.9) $\hat{f}_{t|t} = \hat{f}_{t|t-1} + P_{t|t-1} F_t' [x_t - \hat{f}_{t|t-1}]$ with
     $F_t = P_{t|t-1} + \Phi$,
     $P_{t|t-1} = B_t P_{t-1|t-1} B_t' + \Omega$, and
     $P_{t|t} = P_{t|t-1} - P_{t|t-1} B_t' F_t^{-1} B_t P_{t|t-1}$.

Here unknown parameters are Φ and Ω. In BFP (1990), $t = 0$ is chosen as the opening time of the 16-th (Friday) with $P_{0|0} = I$, x_0 is as observed on the day, Ω is estimated based on the data of Friday, and $\Phi = 0.005 I$. The period (I) between the closing time of Friday and the opening time of Monday and the period (II) between the closing time of Monday and the opening time of Tuesday are treated in two ways; (i) $I = II = 0$ and (ii) $II = 1$ hour, 2 hours, and 17.5 hours.

 On the basis of the estimates (filters) in (3.9) we can estimate the MMI as

(3.10) $\hat{I}_t = (\hat{f}_{t|t}' 1)/d$ $1 = (1, \cdots, 1)': 20 \times 1$

     $\text{Var}(\hat{I}_t) = 1' P_{t|t} 1/d^2$

where d is the denominator in the MMI. The 2 sigma confidence
interval of the MMI_t and its future index are drawn in Figure 3-1.

Fig. 3-1   2-sigma confidence interval

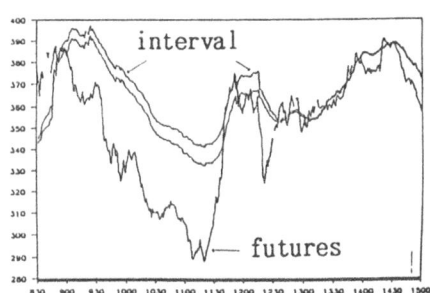

October 20th, 1987

4   Some special multifactor models

A. Factor analysis model

In association with the APT (arbitrage pricing theory), a clas-
sical factor analysis (FA) model is often used as a multifactor
model in (1.1) where factors are unobservable and latent variables.
A typical FA model with means subtracted is described as

$$(4.1) \quad x_t = \beta_1 g_{1t} + \cdots + \beta_q g_{qt} + \varepsilon_t$$
$$= B_0 g_t + \varepsilon_t, \quad \text{where}$$

$$x_t = (x_{1t}, \cdots, x_{pt})' : p \times 1, \quad \varepsilon_t = (\varepsilon_{1t}, \cdots, \varepsilon_{pt})' : p \times 1,$$
$$B_0 = [\beta_1, \cdots, \beta_q] : p \times q, \quad \text{and} \quad \beta_j = (\beta_{1j}, \cdots, \beta_{pj})' : p \times 1.$$

Assumption of the FA model

(i)   For each j, g_{jt}'s ($t=1,2,\cdots, T$) are iid $N(0,1)$.

(ii) $\{g_{jt}\}$ and $\{g_{kt}\}$ are independent (or uncorrelated) processes
      ($j \neq k$).

(iii) ε_t's are iid $N(0, \Lambda)$ where $\Lambda = \text{diag}\{\lambda_1, \cdots, \lambda_p\}$.

(v) $\{\varepsilon_t\}$ and $\{g_t\}$ are independent (or uncorrelated) processes.

(v) $\beta_j' \beta_k = 0$ ($j \neq k$), and rank(B_0) $= q$.

The assumptions (i) and (ii) imply that g_t's iid $N(0, I)$, while
(iii) implies that ε_{it} and ε_{jt} are independent ($i \neq j$) as Λ is

diagonal. In other words, the common factors f_{jt}'s ($j=1,\cdots,q$) are mutually independent series and are white noise series as time series, while ε_{it}'s are errors specific to each variable x_{it}'s and white noise series as time series, independent of the factors. Therefore,

(4.2) $\{x_t\}$ iid $N(0,\ B_0 B_0' + \Lambda)$,

implying that $\{x_t\}$ is of no time series structure. It is easy to see that this model is a special case of the state space model;

$$f_t \equiv g_t \equiv \eta_t, \quad \{\eta_t\} \text{ iid } N_q(0,\ I) \quad (\text{i.e.,}\ B_t \equiv 0),$$
$$x_t = A_t f_t + \varepsilon_t, \quad \{\varepsilon_t\} \text{ iid } N_p(0,\ \Lambda) \quad \text{with } A_t \equiv B_0,$$

where state variables g_{jt}'s are completely random.

The coefficients β_{ij} of the factors are called factor-loadings. By the assumption, β_{ij} is the covariance of x_{it} and f_{jt} $(=g_{jt})$;

(4.3) $\mathrm{Cov}(x_{it},\ g_{jt}) = \beta_{ij}.$

Hence, the correlation coefficient x_{it} and g_{jt} is

(4.4) $\rho_{ij} = \mathrm{Correl}(x_{it},\ g_{jt}) = \beta_{ij}/(\Sigma_{j=1}^{q}\beta_{ij}^{2}+\lambda_i)^{1/2}$

as $\mathrm{Var}(g_{jt})=1$, and ρ_{ij}^{2} measures the explanatory power of g_{jt} for x_{it}. In particular,

(4.5) $\delta_i(\ell) \equiv \rho_{i1}^{2} + \cdots + \rho_{i\ell}^{2}$

is cumulative explanatory power (cumulative relative contribution) of factors $g_{1t},\cdots,g_{\ell t}$ for x_{it}, which is also often called the commonality of $g_{1t},\cdots,g_{\ell t}$ for x_{it}. If x_{it} is standardized as

(4.6) $z_{it} = x_{it}/(\mathrm{Var}(x_{it}))^{1/2}$

and if the FA model (4.1) is assumed for $\{z_{it}\}$, then

(4.7) $\text{Cov}(z_{it}, g_{jt}) = \text{Correl}(z_{it}, g_{jt}) = \beta_{ij}$

i.e., the factor loading β_{ij} represents the correlation between z_{it} and g_{jt}. Thus when we interpret x_{it}'s as returns of assets, g_{jt}'s as common market factors and ε_{it}'s as errors (risks) specific to each asset, the model is a well-set multifactor model as it stands.   Though returns are assumed to be an iid normal process, which is not the case in reality as has been observed in Chapters 2 and 3, this is a typical situation the APT presuppose as a one-period model and hence the FA model is used as an empirical model to test the APT (see Chapter 7 for an example).

However, since B_0 and Λ are unknown, as a statistical model the FA model carries the following difficulties:

a) The unidentifiability of the parameters: The equation

(4.8) $B_0 B_0' + \Lambda = B_0^* B_0^{*'} + \Lambda^*$

will not give us $B_0 = \pm B_0^*$ and $\Lambda = \Lambda^*$ in general.

b) The unidentifiability of the coefficients and factors: $B_0 g_t = B_0^* g_t^*$ allows the indeterminacy with respect to orthogonal transformations as $B_0^* = B_0 \Gamma$ and $g_t^* = \Gamma' g_t$ yields $B_0 g_t = B_0^* g_t^*$ where Γ is a $q \times q$ orthogonal matrix.

c) There may not exist the unique maximum of the log-likelihood equation.

A sufficient condition for the identifiability of (4.8) is given by Tsumura and Fukutomi (1973) and for the problem c) the readers may be referred to Tsumura and Sato (1978).   The indeterminacy cannot be overcome so long as the assumption (i) is kept.   If we replace it by

g_{jt}'s iid $N(0, \nu_j)$ with $\nu_1 > \cdots > \nu_q$,

then at least b) itself diminishes up to the indeterminacy of simultaneous sign changes $B_0 = \pm B_0^*$ and $g_t = \pm g_t^*$.

The basic problem associated with these difficulties will lie in the specification of the model. In fact, although there are only p variables x_{it}'s in the left side of (4.1) at each t, there are $p+q$ variables g_{jt}'s and ε_{it}'s. This means that x_t and $(g_t,$ $\varepsilon_t)$ are not in one-one correspondence and hence for given x_t, g_t and ε_t are not in general identified. To see this point further, let B_1 be a $p \times (p-q)$ matrix such that

(4.9) $\text{rank}(B_1) = p-q$ and $C_0 \Lambda C_1' = 0$ with

(4.10) $[B_0, \ B_1]^{-1} = \begin{bmatrix} C_0 \\ C_1 \end{bmatrix} \begin{matrix} q \\ p-q \end{matrix}$.

Then as $\varepsilon_t = B_0 \varepsilon_{0t} + B_1 \varepsilon_{1t}$ with $\varepsilon_{it} = C_i \varepsilon_t$ ($i=0,1$), the model (4.1) is written as

(4.11) $x_t = B_0[g_t + \varepsilon_{0t}] + B_1 \varepsilon_{1t} = B_0 g_t^* + B_1 \varepsilon_{1t}$

$\qquad = [B_0, B_1] \begin{bmatrix} g_t^* \\ \varepsilon_{1t} \end{bmatrix}$ with $g_t^* = g_t + \varepsilon_{0t}$,

and g_t, ε_{0t} and ε_{1t} are independent with

$\qquad \varepsilon_{0t} \sim N(0, \ C_0 \Lambda C_0')$ and $\varepsilon_{1t} \sim N(0, \ C_1 \Lambda C_1')$.

Of course the choice of B_1 depends on B_0 and Λ and it is not unique. In (4.11)

$\qquad g_t^* = g_t + \varepsilon_{0t} \sim N(0, \ I + C_0 \Lambda C_0'),$

and hence (4.11) is different from the form of the FA model. However, g_{jt}^*'s may be regarded as the common factors affecting x_{it}'s through the factor loading β_{ij}'s. At least, from observable x_t, g_t and ε_{0t} are indistinguishable because from (4.10)

(4.12) $g_t^* = g_t + \varepsilon_{0t} = C_0 x_t,$ $\varepsilon_{1t} = C_1 x_t$ and
$\qquad\qquad g_t^*$ and ε_{1t} are independent.

In other words, g_t^* and ε_{1t} are identifiable from x_t but ε_{0t} and g_t are not distinguishable. It is noted that B_0, B_1, C_0,

C_1 and Λ are related by (4.10) as

(4.13) $B_0 C_0 \Lambda C_0{}' B_0{}' + B_1 C_1 \Lambda C_1{}' B_1{}' \equiv \Lambda.$

Finally it is emphasized that the distinction between common factors g_{jt}'s and specific errors ε_{it}'s simply lies in the fact that g_{jt}'s are supposed to affect x_{it}'s commonly, not in the specification of the processes. All the processes of g_{jt}'s and ε_{it}'s are mutually independent white noises. Letting

(4.14) $A_0 = B_0 D^{-1/2}$ and $f_t = D^{1/2} g_t$ with
$\qquad B_0{}' B_0 = D = \mathrm{diag}(\delta_1, \cdots, \delta_q),$

one may write the model (4.1) as

(4.15) $x_t = \alpha_1 f_{1t} + \cdots + \alpha_q f_{qt} + \xi_1 \varepsilon_{1t} + \cdots + \xi_p \varepsilon_{pt},$

where $A_0 = [\alpha_1, \cdots, \alpha_q]$ and $\xi_i = (0, \cdots, 0, 1, 0 \cdots 0)'$. Here $\{f_{jt}\}$ iid $N(0, \delta_i)$. This expression corresponds to that of the principal component analysis model though (4.15) contains a redundancy because there are $p + q$ components in the right side. In (4.15), if α_k is close to ξ_j for some k and j there is no need to distinguish f_{kt} and ε_{jt}.

An application of a factor analysis model to Japanese stock market is given in Chapter 7.

B. Pena=Box's multifactor time series model

Pena and Box (1987) proposed a time series multifactor model which may be regarded as a time series extension of the FA model. In their formulation there exist q common time series factors (variance components) such that x_t is expressed as

(4.16) $x_t = \alpha_1 f_{1t} + \cdots + \alpha_q f_{qt} + \varepsilon_t = A_0 f_t + \varepsilon_t,$

where $f_t = (f_{1t}, \cdots, f_{qt})'$ and $A_0 = [\alpha_1, \cdots, \alpha_q]: p \times q$. This

equation is the same as the FA model in its form, but the stochastic specification of f_t and ε_t is different from that of the FA
model. In fact, it is assumed that

(i) $\{\varepsilon_t\}$ is a white noise series with mean 0 and $\text{Cov}(\varepsilon_t) = \Phi$,

(ii) $\{\varepsilon_t\}$ and $\{f_t\}$ are independent processes,

(iii) $\{f_t\}$ follows a VARMA process;

$$f_t = \Pi_1 f_{t-1} + \cdots + \Pi_a f_{t-a} + \eta_t + \Theta_1 \eta_t + \cdots + \Theta_b \eta_{t-b},$$

η_t's iid $N_q(0, \Omega)$, Π_i's and Θ_j's are diagonal, and

(iv) $A_0' A_0 = I_q$.

In (i), Φ need not be diagonal, which is a deviation from the
FA model. But the third feature is the main feature of the model
(4.16) and it allows the model to represent certain time series
phenomena of $\{x_t\}$. Of course, if Π_i's and Θ_j's are all zero,
then f_t's are iid $N_q(0, \Omega)$ and then the model is reduced to a
similar model as the FA model. In fact, in that case

(4.17) x_t's iid $N(0, A_0 \Omega A_0' + \Phi)$,

which is the same as (4.2) if $B_0 = A_0 \Omega^{1/2}$ and Φ is diagonal.
Therefore this model will have the same difficulties a)~c) of the
FA model stated above. In fact, regarding B_i as A_i in (4.9)~(4.
11), we can write x_t as

$$x_t = A_0 [f_t + \varepsilon_{0t}] + A_1 \varepsilon_{1t} \quad \text{with} \quad \varepsilon_{it} = C_i \varepsilon_t \ (i=0, 1).$$

Here again $\{f_t\}$, $\{\varepsilon_{0t}\}$ and $\{\varepsilon_{1t}\}$ are independent and $f_t^* = f_t +$
ε_{0t} follows a VARMA. If $\{f_t\} \sim \text{VARMA}(a, b)$, then $f_t^* \sim \text{VARMA}(a, \max$
$(a, b))$, which is stated in their theorem. Therefore the identifia-
bility of the model will be a problem in this model. It is remarked that as in the case of the FA model p-dimensional x_t is
generated by $p + q$ dimensional components f_{jt}'s and ε_{it}'s.

It is noted that this model distinguishes the common factors

from the error terms by their time series structure though the
VARMA can be a white noise in the special case of $\Pi_i = 0$ and
$\Theta_j = 0$. In the FA model the commmon factors are distinguished
from errors by the common effect of factors g_{jt}'s on x_{it}'s and
the specific effect of errors ε_{it}'s on x_{it}'s, so that the
covariance matrix of ε_t is assumed to be diagonal. In the present
model Φ need not be diagonal and hence the distinguishability of
ε_t and f_t depends heavily on the white noiseness of ε_t and the
time series structure of f_t respectively.

In such a distinction between a factor and an error term, it
occurs in financial time series that the total variance of error
terms dominates the total variance of the systematic parts represented by f_{jt}'s and that it is difficult to model them stably
because of their small variances.

In the next chapter we introduce the MTV (multivariate time
series variance component) model in which factors and errors need
not be distinguished whether or not they may be of time series
structure, because the number of variance components in the MTV
model is assumed to be the same as that of x_{it}'s. In other words,
p variates x_{it}'s are generated by p common variance components
f_{jt}'s, some of which may be regarded as errors specific to some
x's.

Exercises

1. How do you think about the invariance of an economic and financial structure over time?
2. In the AIC or SBIC of (2.11), a model of order m is selected
   when it is minimized. Compare the AIC and SBIC in terms of p,
   m and T.
3. Show that (2.12) is equivalent to (2.14).
4. Show that (3.14) can be put in the form of (3.5) by redefining
   f_t and x_t.

5. Express an ARMA model as a state space model.

6. Extend the Kalman filter algorithm to the case where Ω and Φ depend on time t.

7. In the example B of Section 3, discuss the case with the random walk model in (3.7) replaced by a time series model.

8. Discuss about conditions to overcome the unidentifiability of the parameters in (4.8). In the literature, the indeterminacy of B_0 in b) is more emphasized. However, the problem a) is more serious.

9. Show (4.9), (4.11), (4.12) and (4.13).

CHAPTER 5

MTV MODEL AND ITS APPLICATIONS

1 Introduction

As has been stated, many financial time series are of multivariate covariational structure which evolves gradually and will
not admit a profitable opportunity without risk. Hence this point
must be taken into account in efficiently analyzing such volatile
time series as stock prices, interest rates, exchange rates, etc.
In particular, a multivariate approach is more appropriate to describe these volatile phenomena and a proper length of sample period
is required to be chosen. A typical multivariate time series model
sometimes used in practice is VARMA (vector-valued autoregressive
moving average) model. However, as we discussed in Chapter 4, the
model has the two major difficulties in applications;

(a) overflow of parameters,

(b) lack of identifiable models.

To overcome such difficulties, Kariya (1987) proposed the MTV
(multivariate time series variance component) model as an alternative model for the analysis and prediction of multidimensional
volatile time series phenomena. It is supposed in the MTV model
that there are a comparatively small number of major common time
series variance components through which the cross-sectional correlation structure and the serial and cross-serial correlation
structure are formed. The MTV model can be regarded as a dynamic
generalization of the PC (principal component) model, but in the
context of multivariate stationary time series theory it is the
simplest time-reversible model so that the spectral density matrix
is real and symmetric, and diagonalized by a certain orthogonal
matrix. But the model itself allows a heteroscedastic nonstation-
arity. Another feature of the MTV model is that the variance
components can follow any nonlinear models such as GARCH (generalized autoregressive conditional heteroscedastic) model, TARCH

(threshold autoregressive conditional heteroscedastic) model, etc. (see Chapter 3).

In Section 2, we characterize a stationary MTV model as a multi-variate stationary process with symmetric autocovariance matrices which commute each other, and review some theoretical features of the model. In Section 4, the model is applied to a structural analysis of monthly averaged Yen-Dollar exchange rates, and in Section 5 to the prediction of 33 monthly Japanese electronics-related stock prices. In Kariya (1991) a diagonostic checking procedure is proposed about whether a given process $\{ x_t \}$ follows an MTV model.

2 Theoretical foundation of MTV model

Suppose that a p-dimensional stochastic process

$$\{ x_t : \ t=0,\pm 1,\pm 2,\cdots \} \quad \text{with} \quad x_t=(x_{1t},\cdots, x_{pt})'$$

is of the structure

$$(2.1) \quad x_t = \mu_t + \alpha_1 f_{1t} + \cdots + \alpha_p f_{pt},$$

where the coefficient vectors $\alpha_j=(\alpha_{1j},\cdots,\alpha_{pj})' : p\times 1$ ($j=1,$ $\cdots, p$) are assumed to satisfy

$$(2.2) \quad \alpha_j' \alpha_k = \delta_{jk} \ (\delta_{jj}=1, \ \delta_{jk}=0 \ \text{if} \ j\neq k)$$

and f_{jt}'s are variance components commonly generating the varia-tions of x_{it}'s. The condition (2.2) is required for the model (2. 1) to be identifiable. Note that from (2.2), $f_{jt}=\alpha_j'(x_t - \mu_t)$ is obtained.

Assumption of MTV model

(A) $F_j=\{ f_{jt} : \ t=0,\pm 1,\cdots \}$ is covariance stationary with mean $E(f_{jt})=0$. The autocovariance of f_{jt} and f_{jt-k} is denoted by

(2.3) $\gamma_j(k) = \text{Cov}(f_{jt}, f_{jt-k})$.

(B) With $\gamma_j \equiv \gamma_j(0) = \text{Var}(f_{jt})$, $\gamma_1 > \gamma_2 > \cdots > \gamma_p > 0$.

(C) F_j and F_k are uncorrelated $(j \neq k)$: $E(f_{jt} f_{ks}) = 0$ for any t, s.

It is noted that no specific models for F_j's are assumed and that they can be nonlinear. In applications we may take for F_j's ARMA, GARCH, Taylor models, etc.

The MTV model can be viewed as a time series extension of a PC (principal component) model. In fact, if (A) is replaced by

(A)' F_j's are white noise series $(j = 1, \cdots, p)$,

then the common components f_{jt} and f_{js} of variation at time t and s are uncorrelated for $t \neq s$. In the MTV model, F_j's follow stationary processes, through which the model can describe the time series correlation structure of variations of $\{x_t\}$ as well as the cross-sectional correlation structure. As stated above F_j's can be nonlinear models such as ARCH and or Taylor models. Hence an MTV model will be compatible with the features of financial returns we observed in Chapter 2, and will be more efficient than a uni-variate model. In fact, taking $\alpha_j = (0, \cdots, 1, 0, \cdots, 0)$ with $\alpha_{jj} = 1$ $(j = 1, \cdots, p)$, it is reduced to p univariate models.

Characterization of MTV model

The MTV model with Assumption is characterized as the simplest multivariate time-reversible model that is generated by p uncorrelated regular stationary processes. To discuss this point, recall that a p-dimensional (weakly) stationary process $\{x_t\}$ is described by its autocovariance matrices $\{\Sigma(k): k = 0, \pm 1, \pm 2, \cdots\}$ where

(2.4) $\Sigma(k) = E(x_t - \mu_t)(x_{t-k} - \mu_{t-k})'$.

It is noted that when $p > 1$, $\Sigma(k)$ is not symmetric but $\Sigma(-k) = \Sigma(k)'$, implying that $\{x_t\}$ is not time-reversible in general. The following theorem characterizes the MTV model.

Theorem 2.1.   A regular stationary process $\{x_t\}$ follows an MTV model if and only if

(a) $\{x_t\}$ is time-reversible, i.e., $\Sigma(-k) = \Sigma(k) = \Sigma(k)'$,

(b) $\Sigma(k)$'s are simultaneously diagonalized by an orthogonal
    matrix, i.e., $\Sigma(k)\Sigma(\ell) = \Sigma(\ell)\Sigma(k)$ for all k and ℓ.

In fact, the MTV model is the simplest time-reversible model that is generated by p uncorrelated regular stationary processes.

The proof is given in Appendix.

It is noted that if $\{x_t\}$ is generated by more than p uncorrelated processes, these uncorrelated processes will not be identifiable from $\{x_t\}$ (see Section 4 of Chapter 4). On the other hand, if $\{x_t\}$ is generated by less than p uncorrelated processes, $\{x_t\}$ will be degenerate.

Properties of MTV model

Now we investigate some other properties of the MTV model.

[1] *Cross-sectional properties.* It follows from (A) that $\mu_t = E(x_t)$. In the sequel, by considering $x_t - \mu_t$ we may assume $\mu_t = 0$. Also by (A) the variances of f_{jt}'s are independent of t; $\gamma_j \equiv \gamma_j(0) \equiv \mathrm{Var}(f_{jt})$ and by (B) $\gamma_1 > \gamma_2 > \cdots > \gamma_p$. Hence by (C) the covariance matrix of x_t in (2.1) is given by

$$(2.5) \quad \Sigma \equiv \Sigma(0) = \sum_{j=1}^{p} \gamma_j \alpha_j \alpha_j' = A \Lambda A',$$

which is independent of t, where

$$(2.6) \quad A = [\alpha_1, \cdots, \alpha_p] : p \times p \quad \text{and} \quad \Lambda = \mathrm{diag}\{\gamma_1, \cdots, \gamma_p\}.$$

Since $A'A = I$ from (2.2), (2.5) means that γ_j's are the latent roots of Σ and α_j's are the latent vectors corresponding to the γ_j's. This implies that the MTV model cross-sectionally (or for each t) corresponds to the PC model. However, our model is of the time series structure discussed above.

[2] *Time series properties.* To study the time series structure, set $t=1,\cdots, T$ by the stationarity without loss of generality and let

(2.7)
$$y_i=(x_{i1},\cdots, x_{iT})', \quad y=(y_1',\cdots, y_p')' : Tp\times 1,$$
$$f_j=(f_{j1},\cdots, f_{jT})', \quad f=(f_1',\cdots, f_p')' : Tp\times 1.$$

Then the model (2.1) is expressed for $t=1,\cdots, T$ as

(2.8) $y = (A\otimes I) f,$

the covariance matrix of y_i is given by

(2.9) $\text{Cov}(y_i) = \Sigma_{j=1}^{p} \alpha_{ij}^2 \Gamma_j$ with $\Gamma_j=\text{Cov}(f_j): T\times T,$

and the covariance matrix of y is given by

(2.10) $\Omega \equiv \text{Cov}(y) = \Sigma_{j=1}^{p} \alpha_j \alpha_j' \otimes \Gamma_j.$

This covariance matrix describes the whole structure of the model as $\{x_t\}$ is stationary. The (t, s) element of Γ_j in (2.9) is γ_j $(t-s)$ with $\gamma_j(k)$ in (2.3).

[3] *Model identifiability.* Now we shall show that the MTV model formulated above is well-defined. First it is easy to see that if $\Omega=\Omega^*$ for two

$$\Omega = \Sigma_{j=1}^{p} \alpha_j \alpha_j' \otimes \Gamma_j \quad \text{and} \quad \Omega^* = \Sigma_{j=1}^{p} \alpha_j^* \alpha_j^{*'} \otimes \Gamma_j^*$$

satisfying the Assumption, then $\alpha_j=\pm \alpha_j^*$ and $\Gamma_j=\Gamma_j^*$ ($j=1,\cdots, p$). Second if y in (2.8) is expressed in two ways as $y=(A\otimes I) f = (A^*\otimes I) f^*$, then $\Omega=\Omega^*$ implies $\alpha_j=\pm \alpha_j^*$ and $f_j=\pm f_j^*$. Thus except for the simultaneous change of signs of α_j and

f_j the model is identifiable.

[4] *Optimality.*   Next we shall consider an optimality of the
model for the case where, taking q components, we approximate the
variations of x_t ($t=1, \cdots, T$) by the variations of

(2.11) $\hat{x}_t = \alpha_1 f_{1t} + \cdots + \alpha_q f_{qt}$ ($t=1, \cdots T$),

that is, we approximate y by

(2.12) $\hat{y} = (A_q \otimes I)\hat{f}$,

where $A_q = [\alpha_1, \cdots, \alpha_q]$ and $\hat{f} = (f_1', \cdots, f_q')'$.   For an optimality
of the model we make the following assumption.

For any T consecutive $f_{jt} \in F_j$ ($t = r+1, \cdots, r+T$), the
covariance matrices of $f_j(r) = (f_{j,r+1}, \cdots, f_{j,r+T})' : T \times 1$ ($j=1,$
$\cdots, p$) satisfies in terms of positive definiteness

$$\mathrm{Cov}(f_1(r)) > \mathrm{Cov}(f_2(r)) > \cdots > \mathrm{Cov}(f_p(r)).$$

Theorem 2.2.   (Kariya (1987))   Fix q arbitrarily.   Consider the
problem of approximating $y : Tp \times 1$ in (2.8) by a random vector
$y^* : Tp \times 1$ of the form

(2.13) $y^* = (C_q \otimes I)w$,

that is, the problem of approximating $y_i : T \times 1$ by

(2.14) $y_i^* = \Sigma_{j=1}^p c_{i,j} w_j$ ($i=1, \cdots, p$),

where $C_q = (c_{i,j})$ is any $p \times q$ constant matrix satisfying $C_q' C_q = I_q$, and $w = (w_1', w_2', \cdots, w_q')' : Tq \times 1$ with $w_j : T \times 1$ having
$E_j \equiv \mathrm{Cov}(w_j)$ is any random vector satisfying

$$E_1 > E_2 > \cdots > E_q \quad \text{and} \quad \mathrm{Cov}(w_j, w_k) = 0 \ (j \neq k).$$

Then the unique random vector y^* that minimizes in terms of nonnegative definiteness the $T \times T$ mean square matrix

(2.16) $\Phi = \sum_{i=1}^{p} E(y_i - y_i^*)(y_i - y_i^*)'$

is $\hat{y}$ in (2.12). And the minimum value of Φ is $\sum_{j=q+1}^{p} \Gamma_{.j}$.

Corollary 2.1. In the problem of Theorem 2.1, the unique random vector y^* which minimizes $E\|y - y^*\|^2$ is $\hat{y}$ in (2.12).

Corollary 2.2. In the problem of approximating x_t by a random vector of the form $x_t^* \equiv C_q z_t$ where $C_q: p \times q$, $C_q' C_q = I$ and z_t is an arbitrary vector with $Cov(z_t)$, $\hat{x}_t$ in (2.11) minimizes $E\|x_t - x_t^*\|^2$.

The result in Corollary 2.2 and Theorem 2.2 respectively correspond to those of Darroch (1965) and Okamoto and Kanazawa (1968) *in the PC model*, and our results here are regarded as generalizations of those.

[5] *Degree of approximation.* Theorem 2.2 means that the best approximation to a phenomenon following an MTV structure is obtained by taking the q components corresponding to the first q largest covariance matrices in (B) and forming (2.12). A degree of the approximation may be measured by the relative ratio

(2.17) $\eta_q = trCov(\hat{y})/trCov(y) = \sum_{j=1}^{q} \gamma_{.j} / \sum_{j=1}^{p} \gamma_{.j}$.

Here note that $trCov(\hat{y}) = tr \sum_{j=1}^{q} \alpha_j \alpha_j' \otimes \Gamma_j = T \sum_{j=1}^{q} \gamma_{.j}$, since the diagonal elements of Γ_j are all $\gamma_{.j}$. The measure in (2.17) is nothing but the rate of cumulative contribution in the PC model. It is noted that when x_t's are approximated by $\hat{x}_t$'s in (2.11), then the residuals $(x_t - \hat{x}_t)$'s still exhibit a stationary process. Hence there remains some information to exploit in forecasting future values of x_t's. However, in applications we consider the model itself an approximation to a given phenomenon to be analyzed.

[6] *Heteroscedastic MTV model.* When the variances of f_{jt}'s depend

on t, $\mathrm{Cov}(\boldsymbol{x}_t)$ is given by $\boldsymbol{\Sigma}_t = A\Lambda_t A'$ with $\Lambda_t = \mathrm{diag}\{\gamma_{1t}, \cdots,$
$\gamma_{pt}\}$ where $\gamma_{1t} > \cdots > \gamma_{pt}$, and $\mathrm{Cov}(\boldsymbol{y})$ in (2.7) becomes

$$\Omega_T = \Sigma_{j=1}^{p} \boldsymbol{\alpha}_j \boldsymbol{\alpha}_j' \otimes \Gamma_{jT} \quad \text{with} \quad \Gamma_{jT} = \mathrm{Cov}(\boldsymbol{f}_{jT}),$$

where Γ_{jT} depends on the choice of $\boldsymbol{f}_{jT} = (f_{j1}, \cdots, f_{jT})'$ in (2.7).
Then though Theorem 2.1 no longer holds, Corollary 2.2 holds. To
see the estimability of the parameter $\alpha_{i,j}$'s, note the fact

$$E(\boldsymbol{X}\boldsymbol{X}') \;=\; \Sigma_{t=1}^{T} E(\boldsymbol{x}_t \boldsymbol{x}_t') \;=\; A(\Sigma_{t=1}^{T} \Lambda_t) A',$$

where $\boldsymbol{X} = [\boldsymbol{x}_1, \cdots, \boldsymbol{x}_T]$. Hence from this relation an orthogonal
matrix diagonalizing $\boldsymbol{X}\boldsymbol{X}'$ may be regarded as an estimate of A and
the i-th latent root d_i of $\boldsymbol{X}\boldsymbol{X}'$ as an estimate of $\Sigma_{t=1}^{T}\gamma_{jt}$.
This viewpoint facilitates or validates the PC technique applied to
time series data with trends.

3 How to use the MTV model and estimation

The MTV model aims in part to find, through the time series
behaviors of variance components and the sizes of their coefficients, what the time series variance components are. For this
purpose, we may assume for F_j's

(3.1) $\{f_{jt}\}$'s $\sim$ ARMA, ARCH, Taylor models, etc.,

and estimate the parameters of the models based on the scores of
f_{jt}'s. Of course, the identification of the components often
tends to be subjective. In the sequel we list some typical ways of
using the model. For easier understanding, the exposition is also
made in terms of the standardized variables

(3.2) $z_{it} = (x_{it} - \mu_{it})/\sigma_{ii}^{1/2}$,

in which case we assume that z_{it}'s follow the MTV process (2.1).
[1] *Analysis of Variance.* Suppose the first q components are

chosen based on the measure of the cumulative relative contribution in (2.17) as an approximation. Then the size (variance) of the variations of each x_{it} is approximated as

$$(3.3) \quad \mathrm{Var}(x_{it}) \simeq \alpha_{i1}^2 \gamma_1 + \cdots + \alpha_{iq}^2 \gamma_q,$$

where $\gamma_j = \mathrm{Var}(f_{jt})$ as before. Hence the relative explanatory power of the j-th component f_{jt} for the total variations of x_{it} is given by

$$(3.4) \quad \mathrm{Var}(\alpha_{ij} f_{jt})/\mathrm{Var}(x_{it}) = \alpha_{ij}^2 \gamma_j / \sigma_{ii} \equiv \beta_{ij}^2.$$

In other words, (3.3) gives the decomposition of the variance σ_{ii} into the component variances $\alpha_{ij}^2 \gamma_j$ ($j = 1, \cdots, q$). Further β_{ij}^2 is equal to the squares of the correlation coefficient between the i-th variable x_{it} and the j-th component as

$$(3.5) \quad \mathrm{Correl}(x_{it}, f_{jt}) = \mathrm{Cov}(x_{it}, f_{jt})/[\mathrm{Var}(x_{it})\mathrm{Var}(f_{jt})]^{1/2}$$
$$= \alpha_{ij} \gamma_j / (\sigma_{ii} \gamma_j)^{1/2} = \alpha_{ij} \gamma_j^{1/2} / \sigma_{ii}^{1/2} \equiv \beta_{ij}.$$

Hence β_{ij} is a measure of the effect of f_{jt} on x_{it} and it is sometimes referred to as PC factor loading.

In the case of the standardized variables z_{it} in (3.2), the variance of each variable z_{it} is decomposed as

$$(3.6) \quad \mathrm{Var}(z_{it}) = 1 \simeq \alpha_{i1}^2 \gamma_1 + \cdots + \alpha_{iq}^2 \gamma_q,$$

through which the dependence of the variance of z_{it} on that of the j-th component is found out to be $\alpha_{ij}^2 \gamma_j$. Hence in this case

$$(3.7) \quad \mathrm{Correl}(z_{it}, f_{jt}) = \alpha_{ij} \gamma_j^{1/2} = \beta_{ij}.$$

In terms of β_{ij}'s, the model for z_{it} is given as

$$(3.8) \quad z_{it} = \beta_{i1} g_{1t} + \cdots + \beta_{iq} g_{qt} + \varepsilon_{it},$$

where $g_{jt} = f_{jt}/\gamma_j^{1/2}$. Here g_{jt} is often called a factor score while f_{jt} is called a PC score.   Note Var(g_{jt})=1.

[2]   *Cross-Sectional Correlation*.   For each fixed t, the cross-sectional covariance between x_{it} and x_{kt} is decomposed as

$$(3.9)\quad \text{Cov}(x_{it},\ x_{kt}) \simeq \alpha_{i1}\alpha_{k1}\gamma_1 + \cdots + \alpha_{iq}\alpha_{kq}\gamma_q.$$

Hence the cross-sectional correlation between x_{it} and x_{kt} is given by

$$(3.10)\quad \text{Correl}(x_{it},\ x_{kt})$$
$$\simeq [\alpha_{i1}\alpha_{k1}\gamma_1 + \cdots + \alpha_{iq}\alpha_{kq}\gamma_q]/(\sigma_{ii}\sigma_{kk})^{1/2}$$
$$= \beta_{i1}\beta_{k1} + \cdots + \beta_{iq}\beta_{kq} \quad \text{with} \quad \beta_{ij} \text{ in } (3.5),$$

through which we can understand how the cross-sectional correlation structure of x_{it} and x_{kt} is formed through the common components f_{jt}'s.

In the case of the standardized variables z_{it}'s in (3.2), the cross-sectional covariance is equal to the correlation between z_{it} and z_{jt} and it is decomposed as

$$(3.11)\quad \text{Correl}(z_{it},\ z_{jt}) \simeq \alpha_{i1}\alpha_{j1}\gamma_1 + \cdots + \alpha_{iq}\alpha_{jq}\gamma_q,$$

through which it can be established how much z_{it} and z_{jt} are correlated through the k-th component; $\alpha_{ik}\alpha_{jk}\gamma_k$.

[3]   *Time Series Correlation*.   A feature of the MTV model is that the time series covariance structure of x_{it} and $x_{kt-\ell}$ is described by the time series structures of the common components f_{jt}'s as

$$\text{Cov}(x_{it},\ x_{kt-\ell}) \simeq \alpha_{i1}\alpha_{k1}\gamma_1(\ell) + \cdots + \alpha_{iq}\alpha_{kq}\gamma_q(\ell).$$

Also the cross-serial correlation between z_{it} and $z_{jt-\ell}$ is evaluated as

$$\text{Correl}(z_{it},\ z_{jt-\ell}) \simeq \alpha_{i1}\alpha_{j1}\gamma_1(\ell) + \cdots + \alpha_{iq}\alpha_{jq}\gamma_q(\ell)$$

where $\gamma_k(\ell) = \text{Cov}(f_{kt},\ f_{kt-\ell})$ is the autocovariance function of

$\{f_{kt}\}$ with $\gamma_k(0) \equiv \gamma_k$. Hence if $\alpha_{i,j}$'s and $\gamma_k(\ell)$'s are estimated, an estimate of (3.12) is obtained by substitution. For example, when an ARMA model is assumed for $F_k = \{f_{kt}\}$, $\gamma_k(\ell)$ is obtained. Here it is noted that in our model the correlation between z_{it} and $z_{jt-\ell}$ is equal to the correlation between $z_{it-\ell}$ and z_{jt}. Hence a causal relation is not directly observed through the model. However, if it is required, by adding z_{jt-1}, z_{jt-2}, etc. to the set of variables z_{it}'s in the model it will be analyzed.

[4] *Prediction.* Fitting a time series model to $\{f_{jt}\}$ as in (3.1) makes it possible to forecast the ℓ period ahead value $f_{kt+\ell}$ of f_{kt} ($k=1, \cdots, q$). Hence from $z_{it} \simeq \alpha_{i1} f_{1t} + \cdots + \alpha_{iq} f_{qt}$, the future values of z_{it}'s can be forecasted.

Naturally behind this predictability it is necessary for the coefficients α_{ij}'s and the time series models to be stable. When we wish to predict future values of a specific variable, say z_{1t} (such as exchange rate), then it is required to choose those components which explain the variation of z_{1t} well up to a certain level, say 95%.

Estimation

It was observed in Section 2 that the coefficient vector α_i is a latent vector of the covariance matrix Σ of x_t. Hence α_i can be estimated based on the sample covariance matrix;

$$S = \frac{1}{T}(X - M)(X - M)' = \frac{1}{T}\Sigma_{t=1}^{T}(x_t - \mu_t)(x_t - \mu_t)',$$

where $X = (x_1, \cdots, x_T): p \times T$ and $M = (\mu_1, \cdots, \mu_T)$. Here the trends μ_t's are assumed to be known though they must be estimated in the usual manner. As has been seen, $E(S) = \Sigma = A \Lambda A'$ and so by a view of the moment method A may be estimated as an orthogonal matrix which diagonalizes S. In other words, we can estimate α_i's and f_{jt}'s by the PC method as follows.

(1) Estimate γ_i by d_i and α_i by h_i where d_i is the i-th

largest latent root of S and h_i the corresponding latent vector, and

(2) estimate f_{jt} by $\hat{f}_{jt} = h_j'(x_t - \mu_t)$.

We shall show this PC method is in fact the least squares (LS) method. By Corollary 2.1 the best equation which approximates the variation of $V \equiv X - M$ in terms of the MSE is given by $A_q A_q' V$. Therefore we shall try to find an estimate $\hat{A}_q$ which minimizes the squared error

$$\delta = \| V - A_q A_q' V \|^2 = \mathrm{tr}(V - A_q A_q' V)(V - A_q A_q' V)',$$

which becomes the MSE when its expectation is taken. Writing $S = HDH'$, $H'H = I_p$ and $D = \mathrm{diag}\{d_1, \cdots, d_p\}$ where $d_1 > \cdots > d_p$, then

$$\frac{1}{T}\delta = \mathrm{tr}\, D - \mathrm{tr}\, A_q' HDH' A_q = \mathrm{tr}\, D - \mathrm{tr}\, B' DB$$

with $B = H' A_q$. Here noting $B'B = I_q$ (and arguing as in the proof of Theorem 2.2), δ is minimized when $B = (B_1', 0')'$ with $B_1 = \mathrm{diag}\{\pm 1, \cdots, \pm 1\}$ (signs are arbitrary). Hence $\hat{A}_q = (\pm h_1, \cdots, \pm h_q)$ is obtained, where h_i is the latent vector of S corresponding to d_i, and then d_i is an estimate of γ_i. Consequently f_{kt} is estimated as $\hat{f}_{kt} = \hat{a}_k'(x_t - \mu_t)$, and based on $\hat{f}_{kt}$'s such models as ARMA, GARCH, TARCH models, etc. are fitted. After models for $\hat{f}_{kt}$'s are identified, the total MTV model may be reestimated by the ML (maximum likelihood) method when the distributions of $\{f_{jt}\}$'s are available. But the MLE's may not be stable since it is not thought appropriate to take T large and since it is hard to distinguish by data the rather complicated covariance structure of x_t obtained through the autocovariance structure of $\{f_{jt}\}$'s. Here we suggest that the estimates through the PC method and the models based on $\{\hat{f}_{jt}\}$'s be used in practice for stable performance.

4 MTV exchange rate analysis

In this section, a structural analysis of the variations of Yen/ Dollar exchange rates is made via an MTV model. As determinants, we choose the following variables;

(1) monthly average exchange rate (x_{1t}), (2) x_{1t-1},
(3) $x_{1t} - x_{1t-1}$, (4) trade balance*, (5) nontrade balance*,
(6) domestic accumulated capital balance*, (7) foreign accumulated capital balance*, (8) accumulated current balance*,
(9) long-term capital balance*, (10) export contract volume**,
(11) import contract volume**, (12) export amount by letter of
credit**, (13) US short-term prime rate, (14) Japan short-term
interest (gensaki) rate, (15) US TB rate, (16) US-Japan real
interest rate difference, (17) US Federal Fund rate,
(18) Nikkei Stock Index, (19) US Stock Index
                            *: 6 month average, **: 3 month average

These variables will be classified as

(a) fundamental factors: (4)~(12)
(b) factors of US & Japan interest and inflation: (13)~(16)
(c) policy factor: (17)
(d) expectations factor: (3)
(e) stock market factor: (18), (19).

The differenced variable (3) is introduced to reflect such an expec- tations factor as a Band Wagon effect that a conceivable apprecia- tion of Yen moves it up further through the expectations of a further appreciation, while the lagged variable (2) is supposed to reveal the time series structure of the MTV model more clearly.

For the variables (1)~(19), the monthly data is used for the period 1978.1~1985.8 and standardized as $z_{it} = (x_{it} - \overline{x}_i)/s_i$. In the expression of factor scores,

$$z_{it} = \beta_{i1} g_{1t} + \beta_{i2} g_{2t} + \cdots + \beta_{ip} g_{pt} \quad \text{with}$$
$$\beta_{ij} = \alpha_{ij} \sqrt{\gamma_j(0)}, \quad g_{jt} = f_{jt}/\sqrt{\gamma_j(0)}.$$

Table 4-1 lists the variances ($\gamma_j(0)$) and relative contribution rates of f_{jt}'s and the cumulative contribution rates ($j=1, \cdots 5$). Further it lists the factor loading coefficients β_{ij}'s of the ex- change rate variable z_{1t}, the relative contribution rate $\beta_{1j}^2 \times 100$

(%) of each f_{jt} to the exchange rate z_{1t}, and the cumulative explanatory power $100 \sum_{k=1}^{j} \beta_{1k}^2$.

It is observed from this table that in order to grasp the variations of the 19 variables more than 90%, it is necessary to take

Table 4-1

j	①	②	③	④	⑤
variance (γ_j)	8.68	4.89	1.99	1.16	0.77
relative contribution (%)	45.7	25.7	10.5	6.1	4.1
cumulative contribution (%)	45.7	71.5	81.9	88.0	92.1
factor loadings β_{1j} of z_{1t}	0.64	-0.23	-0.54	-0.43	-0.07
$100 \times \beta_{1j}^2$	40.8	5.2	29.0	18.3	0.5
$100 \times \sum_{k=1}^{j} \beta_{1k}^2$	40.8	46.0	75.0	93.3	93.7

the five components, but that it suffices to take the first four components for the explanation of the variations of the exchange rates. In fact, the relative contribution rate of the fifth component f_{5t} to the exchange rates is less than 1% and the cumulative contribution rate of the first four components attains 93%. Consequently we just pay our attention to the ①～④ components.

It is noted that the relative contributions of these components to the exchange rates are not in the order of the number of the components. The component which explains the variations of z_{1t} most is the first component (41%), the third component is ranked in the second (29%), the fourth component in the third (18%) and notably the second component in the last (5%). The graphs of these components are given in Figure 4-1 and the factor loading coefficients of the 19 variables with respect to these 4 components are listed in Table 4-2. It is noted that the factor loading coefficient β_{ij} denotes the correlation coefficient between z_{it} and g_{jt}. In the sequel each component is discussed.

① The first component f_{1t}.

Applying an ARMA model to $\{ f_{1t} \}$, the AR(1) model

$$f_{1t} = 0.9640\, f_{1t-1} + \varepsilon_{1t}$$

is selected based on the AIC criterion. The reason why the graph of

Fig. 4-1 Graphs of variance components

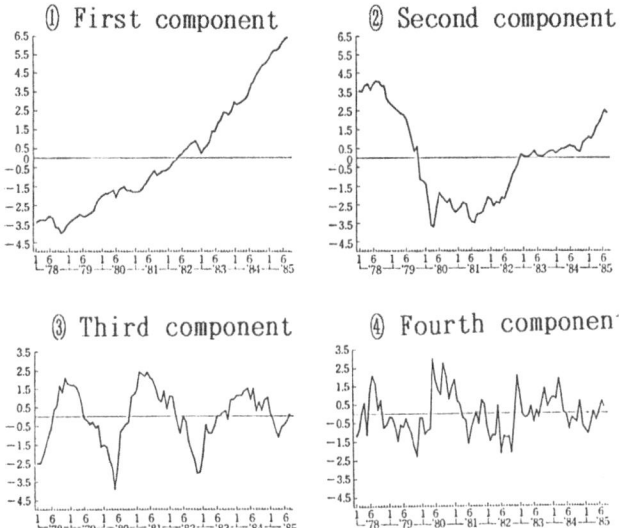

① First component ② Second component

③ Third component ④ Fourth component

Table 4-2 Factor loading coefficients

	①	②	③	④
(1) exchange rate x_{1t}	0.6389	-0.2283	-0.5381	-0.4275
(2) x_{1t-1}	0.6344	-0.1595	-0.6808	-0.1212
(3) $x_{1t} - x_{1t-1}$	0.0101	-0.1828	0.3824	-0.8165
(4) trade balance	0.7706	0.4589	0.2663	-0.0141
(5) nontrade balance	0.3454	0.8603	-0.2095	-0.0339
(6) domestic accum. cap. bal.	0.9941	0.0336	0.0251	-0.0036
(7) foreign accum. cap. bal.	0.9747	-0.0083	0.0855	0.0143
(8) accum. current bal.	0.8096	0.4142	0.0376	-0.0664
(9) long-term cap. bal.	0.8166	0.3308	0.0974	-0.2183
(10) export contract	0.5012	-0.7739	0.1834	0.0660
(11) import contract	0.7064	-0.6517	0.0490	0.1064
(12) export amount by L/C	0.8253	-0.4837	0.2279	0.0865
(13) US short. prime	0.4829	0.1437	0.5214	0.2832
(14) Japan short. interest	-0.0721	-0.7354	-0.3634	0.2617
(15) US TB rate	-0.1982	-0.8017	0.3809	-0.1464
(16) US-J real differ.	-0.5359	0.5881	0.4877	-0.0407
(17) US FFR	-0.3282	-0.8582	0.2165	-0.1150
(18) Nikkei	0.9746	0.0171	0.1177	0.0543
(19) US DJ Index	0.9196	-0.0540	0.1318	0.1780

$\{f_{1t}\}$ in Figure 4-1 is rapidly increasing in spite of the fact
that the coefficient of this equation is less than 1 is that modelling is based on the data (f_{1t}, f_{1t-1}) ($t = 2, \cdots, T$), while the
graph is based on the data (t, f_{1t}) ($t = 1, \cdots, T$). The variables
which have a big correlation (factor loading coefficient) with the
first component are (6) domestic accumulated capital balance (0.99),
(7) foreign accumulated capital balance (0.97), (18) Nikkei Index
(0.97), (19) US Stock Index (0.92), etc., which are the variables
of an increasing trend.   Hence the first component will be deeply
associated with this trend movement.   But the more specific economic
reasoning is reserved here.
② The second component f_{2t}.
   Modelling $\{f_{2t}\}$ yields the ARMA(2,1) model:

$$f_{2t} = 1.9604\, f_{2t-1} - 0.9664\, f_{2t-2} + \varepsilon_{2t} + 0.9754\, \varepsilon_{2t-1}$$

The graph of f_{2t} in Figure 4-1 exhibits that it decreased very
rapidly from 1979 to the first half of 1980 and then increased from
the second half of 1981 after the trough of 1980~1981.   Since 1979
is the year of the Second Oil Schock and the effect is considered
to have continued till 1981, this component is judged to have a
strong assosiation with the movement affected by the Second Oil
Schock.   As the variables which have a large correlation in absolute values with the second component, (4) trade balance (0.46) and
(5) nontrade balance (0.86) which are the components of the current
balance have a positive correlation, while (17) US Federal Fund
rate (-0.86), (15) US TB rate (-0.80), (16) US-Japan real interest
difference (-0.59), and (14) Japan short-term interest rate (-0.74)
have a negative correlation.   Since Japan short-term interest rate
moves parallelly with the inflation rate of Japan, this component
will moves reversely with the inflation rate.   Judging from other
business cycle factors, this second component will be interpreted
as the fundamental factor of the economy associated with business

cycles. The variations of the exchange rates depends on those of this factor only up to 5%.

③ The third component f_{3t}.

The third component is modelled as the ARMA(2,1) model:

$$f_{3t}=1.9256\,f_{3t-1}-0.9598\,f_{3t-2}+\varepsilon_{3t}+0.9745\,\varepsilon_{3t-1}$$

Though this model is close to the model of the second component, the graphs of f_{2t} and f_{3t} are quite different. The fluctuations of f_{2t} are rather cyclical. The variables which have a large correlation with this component are interest variables such as (13) US short-term prime rate (-0.86), (15) US TB rate (-0.80), (16) US-Japan real interest difference (-0.59), (14) Japan short-term interest rate (-0.36), etc. The dependency of the exchange rates on this factor is 29%.

④ The fourth component f_{4t}.

This component exhibits a considerably irregular fluctuation as in Figure 4-1. In fact, modelling it yields the ARMA(0,0) model, i.e., the white noise model;

$$f_{4t} = \varepsilon_{4t}.$$

Of course, this component is not predictable. A very interesting fact is that the variables which have a strong correlation with this component are only (1) exchange rate (-0.43) and (3) differenced exchange rate $z_{1t}-z_{1t-1}$. Furtheremore, the relative contribution of the fourth component to the total variation of the 19 variables is 6.1% but the contribution of the component to the variation of the exchange rate is amazingly 18%.

On the basis of these observations, let us make an analysis on the variations of the exchange rate variable z_{1t}. The standardized variable z_{1t} is expressed by Table 4-2 as

$$z_{1t}=0.64\,g_{1t}-0.23\,g_{2t}-0.54\,g_{3t}-0.43\,g_{4t}+\eta_{1t},$$

where $g_{jt} = f_{jt}/\gamma_j^{1/2}$. In this expression, the variance of each
factor g_{jt} is 1 and hence the factors with larger coefficients in
absolute values make a greater effect on z_{1t}. Hence the factors
(components) are ordered as ①, ③, ④ and ② by the sizes of their
effects. As is pointed out above, the fourth component, which is
identified as a white noise, explains $\{z_{1t}\}$ by 18%. Further,
since the variance of the error term η_{1t} is $1-0.93=0.07$, the
variations of the exchange rates depend 25% on the unpredictable
variations (the variance of $-0.43\,g_{4t}$ (0.18) + the variance of η_{1t}
(0.07)=0.25). In other words, 25% of the fluctuations of the
exchange rates is unsystematic and unpredictable. Of course, there
might be a systematic movement in $\{\eta_{1t}\}$, but it will be difficult
to make a stable model for it since the error term contains various
small factors.

Now, let us consider the first component ① which explains the
variation of z_{1t} most up to 41%. The AR(1) coefficient is 0.964,
which is near 1. This may imply that the trend factor is a random
walk factor as the difference $f_{1t} - f_{1t-1} = 0.036\,f_{1t-1} + \varepsilon_{1t}$ may
be approximated by ε_{1t} (see Definition 2.2 in Chapter 2). If it
is the case, the predictability of the exchange rates will diminish
to 34% because 25% of the variations is impossible to predict (100
$-(25+41)=34$). Hence the third component with explanatory power
29% is the main component providing a predictability for the exchange rates and the variables highly correlated to this component
are (13) US short-term prime rate (0.52), (16) US-Japan real
interest difference (0.49), (15) US TB rate (0.38), (14) Japan
short-term interest rate (-0.36), etc. Hence as is discussed in
Section 3, the cross-sectional correlation between the exchange
rate and these variables are formed through this component. For
example, the correlation between (1) and (13) through this component is $0.5214 \times (-0.5381) = -0.28$, implying that a 1% rise in US
TB rate will be likely to depreciate Yen by 0.28%. Also the correlation between (1) and (16) is $0.4877 \times (-0.5381) = -0.26$.

5 Prediction of stock prices

The MTV model described in the previous sections is applied to
the prediction of electronics-related stock prices. The result here
is simply an example and in the practical prediction we need to consider about the problems of trend, the choice of sample period, data
transformation, the choice of stocks, the use of data other than
stock price data, etc. In this example, trends are not removed from
stock prices and the prices are standardized. The stock price series
we adopt here are those of the following heavy electric machinary,
communications equipment and comsumer electronics & parts firms;

(1) Hitachi, (2) Toshiba, (3) Mitsubishi Electric, (4) Yasukawa
Electric Mfg., (5) Shinko Electric, (6) Matsushita Refrigeration,
(7) Matsushita Seiko, (8) Tokyo Electric, (9) Takaoka Electric
Mfg., (10) Nisshin Electric,   (11) NEC, (12) Fujitsu, (13) Oki
Electric Industry, (14) Iwatsu Electric, (15) Nitsuko, (16) Japan
Radio, (17) Matsushita Electric Indsutrial, (18) Sharp, (19)
Anritsu, (20) Kokusai Electric, (21) Sony, (22) TDK, (23) Teikoku
Tsushin Kogyo, (24) Sanyo Electric, (25) Kenwood, (26) Mitsumi
Electric, (27) Alps Electric, (28) Pioneer Electronic, (29) Victor
Co. of Japan, (30) SMK, (31) Yokogawa Electric, (32) Yamatake-
Honeywell, (33) Matsushita Electric Works

For these stock price series, an MTV model analysis is made for
the monthly (end-of-month price) data of the period 1977.1~1984.12
(12X8=96 for each series). The data is adjusted for splits and
standardized as $z_{it}=(x_{it}-\bar{x})/s_i$. From the correlation coefficients of the 33 series, it is observed:

(a) Matsushita Ref.(6), Matsushita Seiko(7) and Takaoka(9) are most-
    ly negatively correlated to the other series.

(b) Kenwood(25) has no large correlation with any other variable.

(c) Except for (6), (7), (9) and (25) above and for Nisshin(10) and
    Mitsumi(26), all the other variables are highly correlated.

(d) Pioneer(28) moves with a moderate correlation with the varia-
    bles except for (6), (7), (9), (10) and (26).

The variations of these 33 series are 96% explained by the first
6 variance components (f_{jt}'s). The variances (eigenvalues) and
the cumulative rate of contribution (the numbers in parenthes) of

the six components are as follows.

① 23 (70%)   ② 12 (81%)   ③ 1.9 (87%)
④ 1.8 (92%)   ⑤ 0.8 (95%)   ⑥ 0.5 (96%)

The correlation (factor loading coefficient) between each variable
and each component and the determination coefficient of each
variable explained by the six components, which is often called
commonality, are listed in Table 5-1.

Table 5-1

	①	②	③	④	⑤	⑥	η (%)
(1)	.965	-.116	.014	.029	.147	.103	97.8
(2)	.953	-.227	-.001	-.126	.040	.047	98.0
(3)	.964	-.061	-.087	-.078	.089	.002	95.4
(4)	.827	-.475	.006	-.116	.162	.107	96.0
(5)	.740	-.325	-.006	-.511	-.026	.004	91.5
(6)	-.495	.764	-.173	-.045	.239	.146	94.0
(7)	-.491	.539	-.312	-.438	.339	-.080	94.2
(8)	.911	.304	.060	.212	-.048	-.077	97.8
(9)	-.258	.565	-.070	-.706	.158	.021	91.4
(10)	.435	.425	.441	-.525	-.363	.027	97.2
(11)	.982	.013	.044	.097	.080	-.019	98.3
(12)	.970	.100	.136	.059	.069	-.020	97.9
(13)	.957	.168	.108	.047	-.005	-.159	98.3
(14)	.933	.180	.032	.188	.177	-.094	97.9
(15)	.899	.261	-.042	.236	.109	-.156	96.9
(16)	.917	.168	.170	-.130	-.022	.125	93.1
(17)	.970	-.012	-.159	.006	.064	-.012	97.1
(18)	.944	-.030	-.052	.116	.180	-.002	94.0
(19)	.900	.311	.277	.000	-.027	.013	98.4
(20)	.893	.304	.279	-.018	-.131	-.015	98.4
(21)	.801	-.429	-.221	-.072	-.194	.181	94.9
(22)	.964	-.073	-.120	.000	.056	.063	95.6
(23)	.800	.486	-.123	.223	.003	.031	94.2
(24)	.839	-.314	-.184	-.280	-.062	-.191	95.6
(25)	-.025	.550	-.648	.210	-.259	.340	95.0
(26)	.578	.747	.144	.081	-.158	-.099	95.4
(27)	.947	.157	-.198	.090	.056	.015	97.2
(28)	.679	.045	-.573	-.183	-.294	-.221	96.1
(29)	.792	-.390	-.380	-.164	.006	-.001	95.1
(30)	.942	-.041	-.226	-.096	.073	.018	95.4
(31)	.919	-.012	.253	-.157	-.020	.208	97.7
(32)	.897	-.086	.320	.034	.117	.177	96.1
(33)	.927	.145	-.148	.148	-.112	-.060	94.0

More specifically, each variable is expressed as

$$z_{it} = \beta_{i1} g_{1t} + \cdots + \beta_{i6} g_{6t} \quad \text{with}$$
$$\beta_{ij} = \alpha_{ij} \sqrt{\gamma_j(0)}, \qquad g_{jt} = f_{jt}/\sqrt{\gamma_j(0)}$$

and β_{ij} (factor loading coefficients) and $\eta_i = \beta_{i1}^2 + \cdots + \beta_{i6}^2$ (commonality) ($i=1,\cdots,33$: $j=1,\cdots,6$) are listed in Table 5-1.
The commonalities of this table show that the six components are jointly capable to explain the variations of each series more than 90%.

Next we model the components f_{jt}'s by an AR model.

① $f_{1t} = 1.2764\, f_{1\,t-1} - 0.2921\, f_{1\,t-2}$
$\quad\quad \bar{R}^2 = 0.96, \quad DW = 1.91, \quad SD = 0.851.$

② $f_{2t} = 1.1642\, f_{2\,t-1} - 0.1964\, f_{2\,t-2} - 0.2248\, f_{2\,t-3} + 0.4354\, f_{2\,t-4}$
$\quad\quad - 0.1220\, f_{2\,t-5} - 0.2591\, f_{2\,t-6} + 0.5093\, f_{2\,t-7} - 0.3548\, f_{2\,t-8}$
$\quad\quad \bar{R}^2 = 0.94, \quad DW = 1.99, \quad SD = 0.466.$

③ $f_{3t} = 0.9632\, f_{3\,t-1} + 0.0521\, f_{3\,t-2} - 0.0338\, f_{3\,t-3} - 0.0675\, f_{3\,t-4}$
$\quad\quad + 0.2089\, f_{3\,t-5} - 0.3158\, f_{3\,t-6}$
$\quad\quad \bar{R}^2 = 0.86, \quad DW = 3.0, \quad SD = 0.563.$

④ $f_{4t} = 0.92156\, f_{4\,t-1}$
$\quad\quad \bar{R}^2 = 0.86, \quad DW = 2.2, \quad SD = 0.462.$

⑤ $f_{5t} = 0.9161\, f_{5\,t-1} + 0.2526\, f_{5\,t-2} - 0.1320\, f_{5\,t-3} - 0.1617\, f_{5\,t-4}$
$\quad\quad + 0.2858\, f_{5\,t-5} - 0.2689\, f_{5\,t-6}$
$\quad\quad \bar{R}^2 = 0.89, \quad DW = 2.0, \quad SD = 0.271.$

⑥ $f_{6t} = 0.9085\, f_{6\,t-1}$
$\quad\quad \bar{R}^2 = 0.79, \quad DW = 1.8, \quad SD = 0.320.$

The model selection is based on the AIC criterion. The time series graphs of the components are drawn in Figure 5-1. In the figures the dotted lines after 1984.12 are the forecasted values for 10 months by the above AR models. Based on these results, let us look at each component.

Fig. 5.1 Graphs of variance components

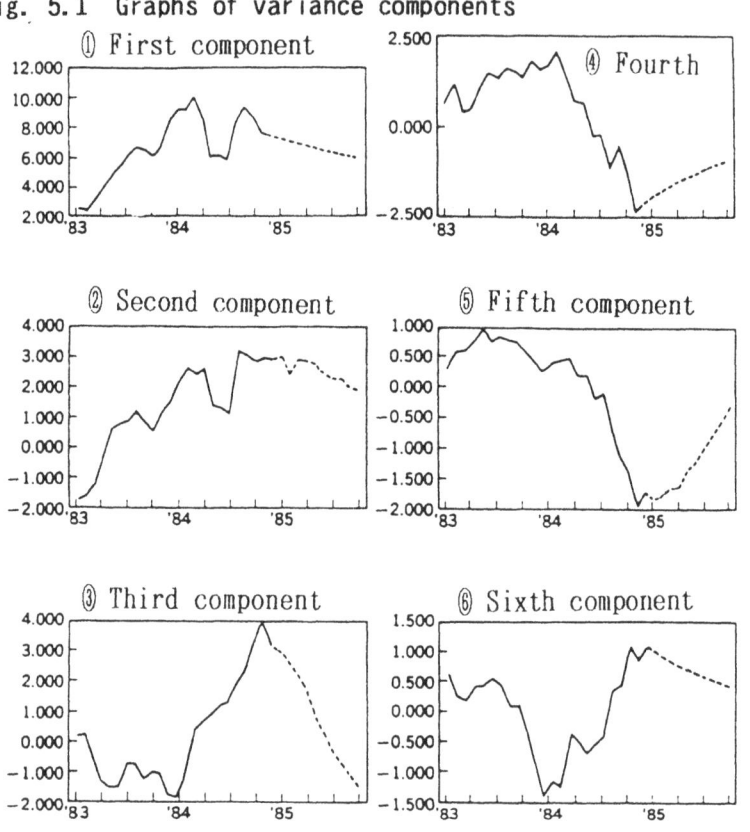

① The first component f_{1t} explains 70% of the total variations of the 33 series.   By Table 5-1, the stocks which depend highly on this component are

(1) Hitach(93%) (2) Toshiba(91%) (3) Mitsubishi(93%) (8) Tokyo (83%) (17) NEC(97%) (12) Fujitsu(94%) (13) Oki(92%) (14) Iwatsu(87%) (15) Nitsuko(81%) (16) Japan R.(84%) (17) Matsu-shita(94%) (18) Sharp(89%) (19) Anritsu(81%) (22) TDk(93%) (27) Alps(90%) (30) SMK(89%) (31) Yokogawa(84%) (32) Y-Honeywell(81%) (33) Matsushita E.W.(86%).

The number in parenthesis is $100\beta_{i1}^{2}$ and shows the determination coefficient (rate of explanation) of the first component ①.   The firms with high $100\beta_{i1}^{2}$ correspond to the internationally renowned general electric and electronic firms and then communications and computer firms follow.   Looking at the graph of ① in Figure 5-1, reflecting the sum $1.2764 - 0.2921 = 0.9843$ of the coefficients of

Fig. 5.2 Graphs of stock prices

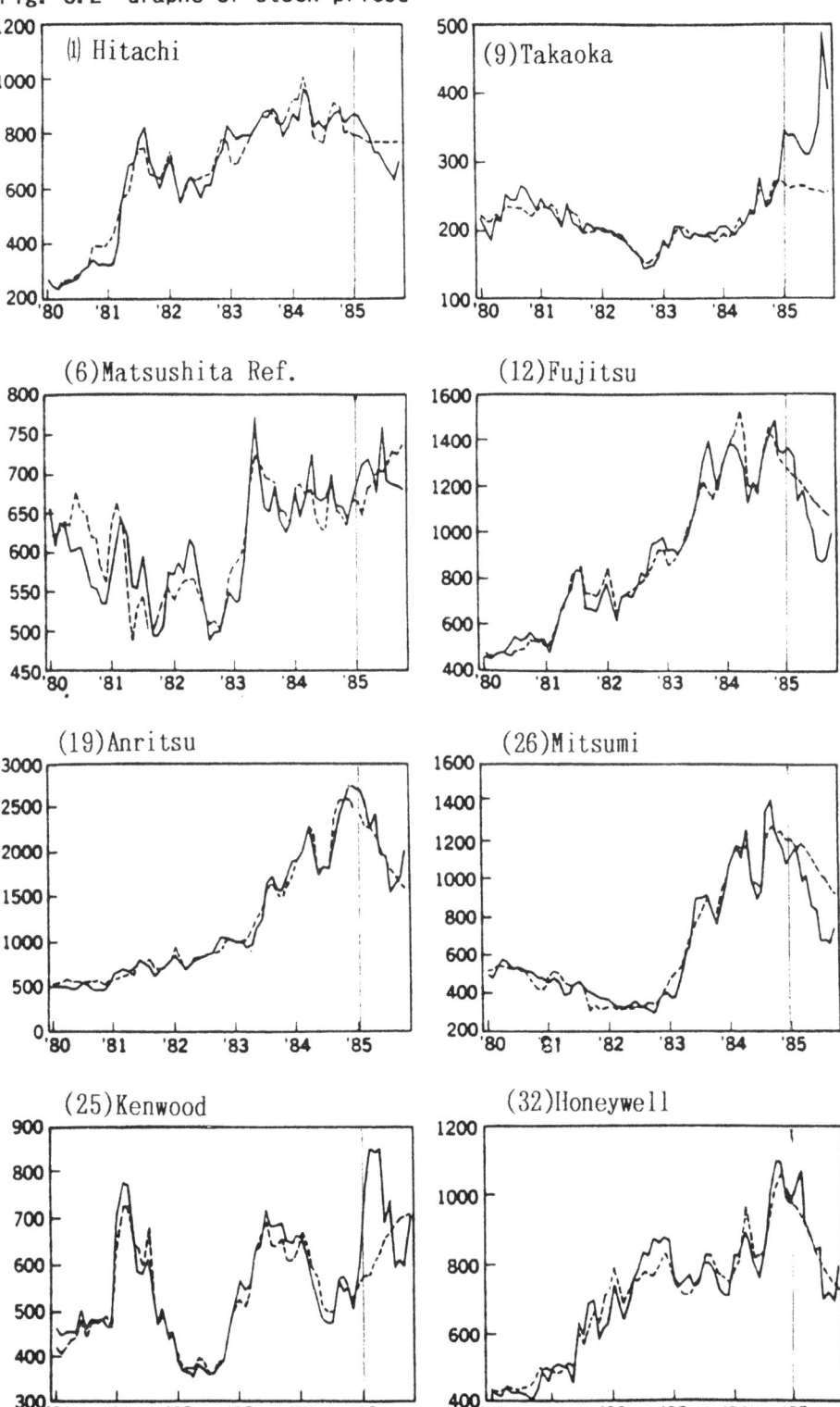

the AR(2) model, the forecasted values show a slowly decreasing
tendency. As examples of forecasts of the stocks belonging to this
group, in Figure 5-2 Hitachi and Fujitsu are taken. In the graphs,
the realized values (solid line) and the interpolated (1977.1~1984.
12) and extrapolated (1985.1~1985.10) values (dotted line) of the
prices are drawn. Needless to say, in the prediction all the six
components are used.

② The second component f_{2t} has a considerably long lag structure
in the model as it is AR(8). The graph of f_{2t} and its predicted
values is given in Figure 5-1 ②. The stocks depending on this component relatively are

(4) Yasukawa(21%)   (6) Matsushita R.(58%)   (7) Matsushita S.(29%)
(9) Takaoka(32%)   (10) Nissin(19%)   (21) Sony(19%)   (23) Teikoku
(24%)   (25) Kenwood(30%)   (26) Mitsumi(56%)   (27) Victor(15%),
etc.

(The number in parenthesis is $100\beta_{i2}^2$)

The interpretation of this component is left out here but the interpolated and extrapolated values of (6) and (26) are graphed in
Figure 5-2.

③ The third component f_{3t} has also a long memory as AR(6). The
stocks which are related to this component are

(10) Nissin(19%)   (25) Kenwood(32%)   (28) Pioneer(33%)
(28) Victor(15%)   etc.

which are the stocks of audio- and micro-chip-related companies.
These series decrease rapidly after 1983, hit the bottom at 1984.11
and recover rapidly again. Hence the graph of f_{3t} in Figure 5-1
should be watched by changing its sign, i.e., as $-f_{3t}$. Taking
this point into account, the graph of f_{3t} in Figure 5-1 predicts
the recovery of its movements. In Figure 5-2 the in-sample and out-
of-sample values of Kenwood(25) is graphed.

④ The fourth component f_{4t} moves as AR(1). Since the coefficient of f_{4t-1} in the model is close to 1, f_{4t} moves like a
random walk. The graph of f_{4t} in Figure 5-1 indicates the recovery after 1984. This component affects the following stocks

(5) Shinko(26%)   (7) Matsushita S.(19%)   (9) Takaoka(50%)
(10) Nisshin(28%),

Table 5-1 picks up the movement of Takaoka 50% of which is determined by the movement of f_{4t}. The predicted values are rather
flat by the effect of the AR(1) prediction and deviated from the
realized values.

⑤ The fifth component f_{5t} is of longer lag structure. But there
is no stock which depends on f_{5t} greatly. Some stocks affected to
some degree by f_{5t} are

(7) Matsushita S.(11%)   (10) Nisshin(13%)   (25) Kenwood(6%)
(28) Pioneer(8%).

⑥ The sixth component f_{6t} is specific to (25) Kenwood(11%).

The forecasting performance of the MTV model in Figure 5-1 seems
encouraging. However in a practical prediction, the combination or
choice of stocks, the transformations of data, the length of sample
period, etc. must be considered carefully, through which the model
performance should be examined.

Exercises

1. Show that if $\{x_t\}$ follows an MTV model, then $\Sigma(-k)=\Sigma(k)=$
   $\Sigma(k)'$ and $\Sigma(k)\Sigma(\ell)=\Sigma(k)\Sigma(\ell)$ for all k and ℓ.

2. Show (2.8), (2.9) and (2.10).

3. Prove the identifiability of an MTV model in [3] of Section 2.

4. Discuss about what Theorem 2.2 implies. In Theorem 2.2, what
   role does w play? (Hint: A main message of Theorem 2.2 is
   that variables should be included in x as far as they are of
   the MTV structure and are objects to be analyzed.)

5. Show (2.17).

6. Analyze financial time series via MTV model in a line with the
   approach in Sections 4 and 5.

7. Show (3.15) and give the details for the procedure of minimizing
   δ in (3.14).

PART II QUANTITATIVE ASSET ALLOCATION SYSTEMS
CHAPTER 6
QUANTITATIVE PORTFOLIO CONSTRUCTION PROCEDURES

1 A comprehensive asset allocation

In general, an investment, whether it may be a financial investment or a real investment, is a commitment to a future world or
future uncertainties and thereby it is inevitably required not only
to forecast the future but also to get involved in possibilities of
risk. Therefore, in decision making for investment, prediction
must be carefully made for evaluation of the future returns and
risks. The forecasting methods for financial investment are often
classified as

(a) methods of fundamental analysis,

(b) methods of technical (chart) analysis,

(c) methods of statistical model analysis, etc.

An important recognition on these methods is that these methods are
not competitive but complementary. In fact, our future world is
too complicated to predict with high precision by a single method.
Behind the fluctuations of financial asset prices, there are many
associated factors we need to take into account and a variety of approaches will reduce the error involved in the prediction of price
variations. However, if a financial investment is an investment
aiming at capital gains as described in Chapter 1, which is a typical case in Japanese stock market, the variations of asset prices
tend to be made in market without a close connection with the fundamentals of our economy or returns from real investment such as
manufacturing investment and then the prediction on the basis of
a value analysis via fundamentals etc. will not function efficiently. In particular, Japanese stocks in which dividends are not
significant for investment are of this feature. The price variations therein are of the game-theoretic uncertainties described in
Chapter 1, and hence statistical and probabilistic model approach
(quants) will become im_portant. In this chapter we develop a

basic theory on financial quants and trading systems from this
viewpoint. It should be recalled that by quants we mean quantitative methods for financial analysis and/or quantitative financial
analysis (see Section 1 of Chapter 1).

A basic form of a portfolio-construction system

In general, a portfolio-construction or an asset allocation
system consists of the following five elements.

(1) Specification of an investment stance about fund of investment,
    target returns, time horizon of investment, risk allowances,
    etc.
(2) Selection of a target population of financial assets for
    investment.
(3) Description or modelling and prediction of price variations.
(4) Construction of a portfolio (optimization).
(5) Choice of a trading (rebalance) rule.

The most important element of the above five is the third element
(3). In fact, if it is possible to model the process of price
fluctuations with considerably high precision of prediction, it
will be also possible to construct a high-performance portfolio.
Of course, it is not easy to find a model with high predictive
power. Especially it will be difficult to predict individual price
variation with high precision. However, it should be noted that
whether or not a price series may be modelled in a statistical manner, investors always predict the future movements of asset prices
in some way, which may be very subjective. Usual approaches to the
third element are, as stated above,
A. fundamentals approach in which asset prices are associated, and
    sometimes statistically modelled, with economic, social and poli-
    tical fundamental factors, then they are forecasted and assets
    for investment are selected,
B. technical approach in which the past trends and patterns of

prices are individually evaluated by an expertized technique and
extrapolated, and

C. quants approach in which the process of prices regarded as a
stochastic process is modelled statistically and the future
values are predicted.

These approaches have a common feature that prediction is made
on the basis of past data, but they differ in processing the information. In approach A,

1) the past and present information (data) on asset prices and
fundamentals, which may not be quantifiable, is made a good use
of on the basis of a subjective judgement (experience, intuition,
knowledge) and of a statistical analysis,

2) a correspondence (model) among the variables is made at least
in analyst's head (personal model), and the future movement of
the fundamentals and prices are forecasted subjectively or
objectively,

3) assets whose prices are forecasted to rise are selected for
investment and

4) a portfolio is formed such as an equally-weighted portfolio.
This portfolio-making procedure is common to the chart approach B
and quants approach C. In B, the price data is mainly used, but
the other parts are similar. In C, the process of prices is statistically modelled with or without fundamentals variables in part 2),
future values of prices are forecasted in 3) and an optimization
technique is usually applied in 4). Therefore the three approaches
are not formally different. But the difference lies in the scope
and treatment of the available information, which gives rise to
differences in performance.

Price variation model

In quants approach C, price variations are modelled statistically based on price data and some other data possibly including funda-

mentals data.   The models may be the ones derived from or used in a
finance theory such as

① CAPM (Capital Asset Pricing Model),

② APT (Arbitrage Pricing Theory) multifactor model, etc.

In the finance theories, these models are usually static, one-
period and cross-sectional model, and statistically they correspond
to multiple regression model, factor analysis model, multivariate
regression model, etc. depending on how the theory is "interpreted
in practice".   However, the process of prices, needless to say, is
a time series phenomenon and therefore modelling it needs to take
into account the time series correlation structure of the process
as well as the cross-sectional correlation.   In addition, it is
also necessary to take into account that a price process will gradually change along with time.

Multivariate time series models often applied or possibly appli-
cable in applications for decision-making in practice are

③ Multivariate time series regression model

④ VAR (vector autoregressive) model

⑤ Linear or Nonlinear MTV model

⑥ Multivariate Nonlinear Taylor

⑦ Nonlinear ARCH model

⑧ Multivariate Regression-MTV mixed model

⑨ Kalman filter model

⑩ Time varying coefficient model

etc.

In modelling a price process with price and fundamentals data in
practice, the length of sample period must be selected properly
relative to an investment time horizon.   If the length is too long
relative to the time horizon of investment, we may suppress the
relevant variations or information into the long-run movement and
simply construct an averaged model over the long period that has
neither an explanatory power near the present time nor a predictive
power.   Even in such a time-varying parameter model treated as a

Kalman filter model, the problem of selecting a proper length of
sample period is important, because the stability of estimates of
fixed parameters in the time-varying parameter model and hence the
model performance are subject to the selection of the length of
sample period. This implies that in modelling a stock price process
the sample period will be relatively short compared to the number
of assets, and hence a subpopulation (group) of assets out of the
whole assets needs to be chosen properly. We call such a subpopulation for portfolios a *target population* or a *portfolio population.*
A selection of a target population of assets may be statistically
made on some variational features corresponding to an investment
stance. Also in that selection, the stability of the variational
features extracted is important.

Goal of quants asset allocation

   In investment, needless to say, financial instruments with high
returns and low risks are preferred. However, it is a general case
that there is a trade-off between returns and risks, as often stated as high return - high risk, and hence we are required to make a
certain concession in investment. Such a concession corresponds to
an investment stance and it is associated with the type or kind of
fund (institutional or governmental or individual, pension or
mutual fund or saving, etc.), time horizon of investment, target
returns, permissible risks, etc. From the viewpoint of a portfolio
quants, such an investment stance corresponds to a specific variational feature of a portfolio to construct and therefore the basic
problem of the portfolio quants is how to stably create the variational features corresponding to a specific investment stance over
a certain period. In other words, the goal of quants asset allocation is to create a new stochastic process via a portfolio which
meets a given investment stance. And an investment stance need to
be formed *after* the variational features of asset prices are under-
stood. Thus the problem is of time series concept rather than of

cross-sectional concept, though the stability of the variational features is heavily dependent on that of the cross-sectional relationship of assets. Further the problem is closely related to the problem of selecting a population of assets for portfolios and constructing an optimal portfolio relative to an investment stance.

A mathematical formulation of the argument made here is given in Section 4.

2 Basic notation and concept

To introduce some basic notation, let $t = T$ be the present time, i.e., the time point when we make an investment decision, let M be the number of assets for investments and let

S_{it} : the price level of the i-th asset at time t,

$S_t = (S_{1t}, \cdots, S_{Mt})'$: $M \times 1$ vector of prices,

b_i : the units of the i-th asset contained in a portfolio,

where $b = (b_1, \cdots, b_M)'$: $M \times 1$ defines a portfolio,

A_T : the total fund to invest at T.

Then the value of a portfolio b at t is

$$(2.1) \quad V_t(b) = b_1 S_{1t} + \cdots + b_M S_{Mt} = b' S_t,$$

where $t = 1, 2, \cdots, T, T+1, T+2, \cdots\cdots,$ and

$$(2.2) \quad V_T(b) = A_T.$$

The time series process $\{V_t(b)\}$ is often called the value process of portfolio b. Assume $b_i \geq 0$ and define

$$(2.3) \quad c_i = \frac{b_i}{\sum_{j=1}^{M} b_j} \quad (\geq 0) \ (i = 1, \cdots, M)$$

with $\sum c_i = 1$. Then c_i is the proportion in units of the i-th asset in the portfolio and

$$(2.4) \quad V_t(c) = c_1 S_{1t} + \cdots + c_M S_{Mt} = c' S_t$$

with $c=(c_1,\cdots,c_M)'$: $M\times1$ is a normalized process of $V_t(b)$;

(2.5) $V_t(c) = V_t(b)/\sum_{j=1}^{M} b_{\cdot j}.$

When b is given, c is obtained by (2.3), while when c is given, then $\Sigma_j\, b_j$ is obtained by

(2.6) $(\sum_{j=1}^{M} b_j)\, V_T(c) = A_T,$

and hence b_i's are obtained by (2.3). Thus considering a portfolio in terms of units b_i's is equivalent to considering it in terms of unit ratios c_i's with initial investment fund A_T at T. It is noted that from (2.5) the return $r_t(b)=\log V_t(b)/V_{t-1}(b)$ of $V_t(b)$ is equal to that of $V_t(c)$. In other words, $r_t(a\,b)=r_t(b)$ for any $a>0$ and hence b can be determined up to the ratios c_i's by an optimizing procedure on the basis of $r_t(b)$'s, which will be discussed later in more details. In terms of returns, it is convenient to treat the problem by the value ratios of the portfolio invested on each asset over the total amount of fund;

(2.7) $d_{iT} = b_i\, S_{iT}/A_T,\quad d_{1T}+\cdots+ d_{MT} = 1.$

For brevity, we call d_{iT}'s *portfolio value ratios* or simply value ratios, and c_i's *portfolio unit ratios* or simply unit ratios. By portfolio ratio we mean either d_{iT} or c_i. Let the t-th price relative to the price S_{iT} at T be

(2.8) $\tilde{x}_{it} = S_{it}/S_{iT}.$

Then $\tilde{x}_{it}-1$ for $t>T$ is the return from present time T to future time t while $\tilde{x}_{it}^{-1}-1$ is the return from past time t to present time T. Using (2.7) and (2.8),

(2.9) $\tilde{x}_t(d_T) = d_{1T}\tilde{x}_{1t}+\cdots+ d_{MT}\tilde{x}_{Mt} = d_T'\,\tilde{x}_t$
$$= V_t(b)/V_T(b) = V_t(b)/A_T,$$

where $d_T = (d_{1T}, \cdots, d_{MT})'$, $\tilde{x}_t = (\tilde{x}_{1t}, \cdots, \tilde{x}_{Mt})'$, and $\tilde{x}_T(d_T)$
$= 1$. Therefore $\tilde{x}_t(d_T) - 1$ is the return of the portfolio from
T to $t > T$. Since we are interested in future returns, the value
ratios d_{iT}'s and the corresponding relative prices $\tilde{x}_{it}$'s in the
form of (2.8) are easy to understand in portfolio analysis (see
Section 3 for details). But a warning is that $\{\tilde{x}_t\}$ will not be
stationary but heteroscedastic.

An asset allocation problem consists of

(a) a selection of a portfolio population of assets for construc-
    ting a portfolio, which may contain such derivative assets as
    futures and options, and

(b) a determination of value ratios d_{iT}'s or unit ratios c_i's
    over the assets chosen to allocate fund A_T.

Theoretically speaking, since $c_i = 0$ or $d_{iT} = 0$ means the dele-
tion of the i-th asset from portfolio, the selection of a portfolio population in (a) is not necessary. However, practically the
part (a) is quite important to construct a reliable or stable portfolio. In fact, as has been discussed, it is aimed in the determination of the portfolio ratios c_i's that a certain variational
characteristic of the value process $\{V_t(c)\}$ is created with
respect to an investment stance. But in creating it, statistically
stable models will be needed and a lot of diagnostic simulations
are needed to carry out. Therefore because of restrictive availa-
bility of data, possibilities of changes of models over time, the
stability of models and a limitation for computational time for the
determination of portfolio ratios, selection of a portfolio population will be required in advance. A selection system of a portfolio
population is often called a screening system.

(a) Methods for selection of a portfolio population

1) On the basis of such financial and accounting data as PBR
   (price book value ratio), PER (price-earning ratio), dividend
   payout ratio, liquidity, etc. assets are screened out by
   certain criteria.

2) Like an index tracking portfolio, stocks or assets are selected
to follow a specified object.

3) Like a concept of "industry", stocks or assets falling into a
promising category are selected.

4) Stocks or assets whose prices are of strong time series structure giving predictability are selected for a stable prediction.

(b) Selection of portfolio ratios

A traditional method of selecting portfolio ratios is the method
of allocating a fund equally to selected assets so that the portfolio value ratios are equal.   But this method does not make a good
use of the correlations of price variations among assets and does
not pay a sufficient attention to the expected risk of the equally-
weighted portfolio though it takes the expected return into account.
In the Markowitz portfolio theory, portfolio ratios are determined
in order to control the future return and risk of portfolios by
using the cross-sectional correlation of asset prices. In practical
optimization, we aim to construct a portfolio by using not only the
cross-sectional correlation but also the time series correlation of
prices, so that the portfolio value process carries a certain variational feature corresponding to an investment stance.

However, even if a portfolio population is given, it will be
necessary to forecast the future value $V_{T+r}(b)$ of a portfolio b
in the determination of portfolio ratios because investment is a
commitment to future.   Hence the distribution of the future value
$V_{T+r}(b)$ of a portfolio b need to be forecasted over $r=1, 2, \cdots$,
based on the processes of asset prices empirically identified and
modelled by a statistical method, and then the portfolio is optimized with respect to the forecasted returns and risks corresponding
to an investment stance.

3 Markowitz Theory and its implications
Markowitz Theory

No doubt, a modern portfolio theory was initiated by Markowitz (1956). A main part of his theory is described as follows. Let

(3.1) $x_{it} = \log S_{it} - \log S_{it-1} = \log(1 + r_{it}) \simeq r_{it}$

where $r_{it} = (S_{it} - S_{it-1})/S_{it-1}$ ($i = 1, \cdots, M$: $t = 1, \cdots, T$). Of course, x_{it} is the continuously compound return of the i-th stock from $t - 1$ to t and r_{it} is the one-period simple return. They are not very different when one-period is not long. In the sequal we discuss our problem in terms of r_{it}'s. Let the means and covariances of r_{it}'s be $E(r_{it}) = \mu_{it}$ and $\mathrm{Cov}(r_{it}, r_{jt}) = \sigma_{ijt}$ ($i, j = 1, \cdots, M$), where $\sigma_{iit} = \mathrm{Var}(r_{it})$, the variance of r_{it}. In general, μ_{it} and σ_{ijt} are dependent on time t. In most theoretical situations, for simplicity it is assumed that they are constant over time:

(3.2) $\mu_{it} \equiv \mu_i$ and $\sigma_{ijt} \equiv \sigma_{ij}$.

The Markowitz theory is a one-period theory with $T = 1$ in which an investor at time 0 is supposed to make an optimal portfolio among the M assets for time $T = 1$ with respect to his preference (objective function) on predicted returns and risks (standard deviations) (This interpretation was in fact due to Tobin). Hence in the (static) Markowitz theory the variability of μ_{it} and σ_{ijt} over time is irrelevant since $T = 1$. Let $d = (d_1, \cdots, d_M)'$ with

(3.3) d_i : portfolio value ratio of the i-th asset at time 0.

Then the portfolio return at time 1 is

(3.4) $v_t(d) = d_1 r_{1t} + \cdots + d_M r_{Mt}$ ($t = 1$),

which has not been realized at time 0. The mean and variance of $v_t(d)$ are respectively given by

$$\mu(d) = \Sigma_{i=1}^{M} d_i \mu_i \quad \text{and}$$

(3.5)

$$\sigma^2(d) = \Sigma_{i=1}^{M} \Sigma_{j=1}^{M} d_i d_j \sigma_{ij}.$$

These are the expected (predicted) return and risk of a portfolio ($d_1, \cdots, d_M$) at time 0. Here if μ_i's and σ_{ij}'s are assumed to be known, then an investor at time 0 is supposed to optimize his objective function

(3.6) $u(\mu(d), \sigma^2(d))$ (e.g. $\alpha \mu(d) - \beta \sigma(d)$, $\alpha, \beta > 0$)

expressing his preference on expected returns and risks. Then the optimal portfolio ratios d_i's are determined at time 0. This is a basic part of the Markowitz Theory.

Portfolio effect

In general, making a portfolio will reduce the total risk of investment, which we shall call the *portfolio effect*.   The portfolio effect consists of the two effects:

(1)   **Markowitz effect** (negative correlation effect),

(2)   **diversification effect** (uncorrelation effect).

A distinction should be made between these two effects because the information we use in data is different.

(1)   It is the principle of the Markowitz effect that if two random variables X and Y are negatively and strongly correlated, the variance of a suitable (positive) combination of X and Y becomes close to zero.   In fact, since

(3.7) $\text{Var}(a X + b Y) = a^2 \sigma_x^2 + 2 a b \sigma_x \sigma_y \rho_{xy} + b^2 \sigma_y^2$
$$= (a \sigma_x - b \sigma_y)^2 + 2 a b \sigma_x \sigma_y (\rho_{xy} + 1),$$

if $a/b = \sigma_y/\sigma_x$ and $\rho_{xy} = -1$, then the variance of $a X + b Y$ is zero, where σ_x^2, σ_y^2 and ρ_{xy} are respectively the variances of X and Y, and the correlation coefficient of X and Y.   In this relation, the number of assets for a portfolio can be small and can be 2 as above. In general, the variance $\sigma^2(\boldsymbol{d})$ of $v_1(\boldsymbol{d})$ in (3.4) can be close to zero with nonnegative value ratios $d_i \geqq 0$ if such a negative correlation among the M assets exists as a whole.

(2) On the other hand, the diversification effect makes use of the uncorrelation structure of the returns of assets. In fact, if random variables $u_1, \cdots, u_p$ are uncorrelated and have the variances $\sigma_1^2, \cdots, \sigma_p^2$ with $|\sigma_i^2| \leq K$ ($i=1, \cdots, p$), then

$$(3.8) \quad \mathrm{Var}(d_1 u_1 + \cdots + d_p u_p) = d_1^2 \sigma_1^2 + \cdots + d_p^2 \sigma_p^2$$
$$\leq (d_1^2 + \cdots + d_p^2) K.$$

Hence if we take $d_i = 1/p$, $\mathrm{Var}(d_1 u_1 + \cdots + d_p u_p) \leq K/p$, which is small if p is large. Therefore, in the diversification effect the number of assets included in a portfolio should be large to reduce the risk of the portfolio.

In practice, we can make a good use of the both effects by distinguishing the "systematic" (negatively correlated) part and "unsystematic" (uncorrelated) part of data, which will be different from the concept customarily treated in such a model as CAPM.

Portfolio value ratio and unit ratio

In a one-period model, the ratios d_i's determined by optimizing (3.6) are the portfolio value ratios relative to the initial investment fund A_0. In fact, if d_i is the value ratio for the i-th asset, then its unit is given by

$$(3.9) \quad b_i = d_i A_0 / S_{i0}$$

(see Section 1). Therefore the values of the portfolio at time 0 and 1 are V_0 and V_1, where

$$(3.10) \quad V_t = \Sigma_{i=1}^M b_i S_{it}$$

($t=0,1$ and $V_0 = A_0$), and the portfolio return is given by

(3.11) $r_t(d) = \dfrac{V_t - V_{t-1}}{V_{t-1}} = \dfrac{\sum_{i=1}^{M} d_i (S_{it} - S_{it-1})/S_{i0}}{\sum_{i=1}^{M} d_i S_{it-1}/S_{i0}}$

though $r_t(d)$ is right now defined only for $t = 1$. Substituting
(3.9) into (3.10) yields $V_0 = A_0 \sum d_i$ and $V_1 = A_0 \sum d_i (S_{i1}/S_{i0})$.
Hence from $\sum d_i = 1$ and $r_{it} = (S_{it}/S_{it-1}) - 1$, the return $r_t(d)$
in (3.11) is equal to $v_t(d)$ in (3.4) for $t = 1$. This implies
that in a one-period model d_i's optimizing (3.6) are portfolio
value ratios and the units b_i's for the corresponding allocation
are determined by (3.9). In other words, the value ratios ($d_1, \cdots,$
d_M) are in one-one correspondence with the units ($b_1, \cdots, b_M$) with
$V_0 = A_0$ in a one-period model.

However, in a multiple-period model this correspondence relation
is no longer true, though in practice it seems to be taken for
granted. To see this in details, suppose that we are at time T
with fund A_T. Then as in (3.9) the value ratio d_{iT} for the i-th
asset is given by

(3.12) $d_{iT} = b_{iT} S_{iT} / A_T,$

where b_{iT} is the unit of the i-th asset in a portfolio. Then the
portfolio return at past time or future time t as defined by (3.11)
is

(3.13) $r_t(d_T) = \dfrac{\sum_{i=1}^{T} d_{iT}(S_{it} - S_{it-1})/S_{iT}}{\sum_{i=1}^{T} d_{iT} S_{it-1}/S_{iT}} = \sum_{i=1}^{T} w_{it-1} r_{it},$

where $r_{it} = (S_{it}/S_{it-1}) - 1$ as before and

(3.14) $w_{it-1} = \dfrac{d_{iT}(S_{it-1}/S_{iT})}{\sum_{j=1}^{M} d_{jT}(S_{jT-1}/S_{jT})}$.

Here w_{it-1}'s depend on t. This implies that the quantity defined
by

(3.15) $v_t(d_T) = d_{1T} r_{1t} + \cdots + d_{MT} r_{Mt}$

(as in (3.4)) is not the portfolio return $r_t(d_T)$ in (3.11) for each t except for $t = T + 1$. Consequently the quantity $v_t(d_T)$ with d_{iT}'s maximizing such an objective function as in (3.6) fails to reflect the return of the portfolio d_T at time t ($t \neq T+1$). In other words, an analysis of the process $\{v_t(d_T)\}$ is not equivalent to an analysis of the process $\{r_t(d_T)\}$ and for an over time optimization or for a study of the returns process of a portfolio d_T one should use $\{r_t(d_T)\}$ rather than $\{v_t(d_T)\}$. As in (3.13) $r_t(d_T)$ is nonlinear in d_{iT}'s though $v_t(d_T)$ in (3.15) is linear in d_{iT}'s.

This observation is very important in applications. In fact, in practice μ_{it}'s and σ_{ijt}'s are unknown and must be estimated by a certain method or model. Then it is often the case that time series data is used and the *quasi* portfolio return process $\{v_t(d_T)\}$ is analyzed over the sample period. But what should be analyzed is the real portfolio return process $\{r_t(d_T)\}$. As an example we shall consider a simplest case.

Historical Markowitz portfolio

To apply the Markowitz theory to past time series data, let us assume that for $t = 1, \cdots, T$ the means μ_{it} and covariances σ_{ijt} are constant as in (3.3). First we describe the method customarily adopted. Since μ_i and σ_{ij} are constant, it is natural to estimate them by the sample mean and covariance:

$$\hat{\mu}_i = \frac{1}{T}\Sigma_{t=1}^T r_{it} \quad \text{and} \quad \hat{\sigma}_{ij} = \frac{1}{T}\Sigma_{t=1}^T (r_{it} - \bar{r}_i)(r_{jt} - \bar{r}_j)$$

and to estimate $\mu(d)$ and $\sigma^2(d)$ in (3.5) we replace μ_i and σ_{ij} by $\hat{\mu}_i$ and $\hat{\sigma}_{ij}$. Then substituting these values into the objective function in (3.6) and optimizing $u(\hat{\mu}(d), \hat{\sigma}^2(d))$ with respect to d, the ratios d_i's obtain. Regarding these ratios as portfolio value ratios at $t = T$, we can get the portfolio unit ratios

(3.16) $c_{iT}^* = A_T d_i / S_{iT},$

where A_T is a fund at T. However the ratios d_i's are not the value ratio except for $t = T+1$. In other words, the portfolio return based on c_{iT}^*'s in (3.16)

$$(3.17) \quad r_t(d) = \frac{\sum_{i=1}^{M} d_i (S_{it} - S_{it-1})/S_{iT}}{\sum_{i=1}^{M} d_i S_{it-1}/S_{iT}}$$

is not equal to $v_t(d) = \sum d_i r_{it}$ and even the average return $\frac{1}{T}\sum_{t=1}^{T} r_t(d)$ is not equal to $\hat{\mu}(d)$. Furthermore $r_t(d)$ in (3.17) is a highly nonlinear function of individual returns r_{it}'s.

To avoid this inconvenience, it will be necessary to discuss the problem in terms of unit ratios which are invariant over time. Let

$$V_t(c) = \sum_{i=1}^{M} c_i S_{it} \quad \text{with} \quad c = (c_1, \cdots, c_M)' \text{ and } \sum c_i = 1$$

be the normalized value of a portfolio at time t (see Section 1). Then the return of the portfolio is given by

$$(3.18) \quad r_t(c) = [V_t(c) - V_{t-1}(c)]/V_{t-1}(c),$$

and the mean and variance of returns over $t = 1, \cdots, T$ are respectively $\hat{\mu}(c) = \frac{1}{T}\sum_{t=1}^{T} r_t(c)$ and $\hat{\sigma}^2(c) = \frac{1}{T}\sum_{t=1}^{T}(r_t(c) - \hat{\mu}(c))^2$. Hence optimizing a given objective function $u(\hat{\mu}(c), \hat{\sigma}^2(c))$, we obtain optimal portfolio unit ratios c_i's, which are defined over time, and hence optimal portfolio value ratios d_{iT}'s at T by $d_{iT} = b_i S_{iT}/A_{0T}$ with $c_i = b_i/\sum b_i$, which depend on T. For example, we may want to maximize

$$(3.19) \quad w(c) \equiv u(\hat{\mu}(c), \hat{\sigma}^2(c)) = \lambda_1 \hat{\mu}(c)^2 + \lambda_2 \hat{\sigma}(c)^2$$

$(\lambda_1 > 0, \lambda_2 < 0)$. Then this is not a quadratic function of c but a highly nonlinear function of c. Thus the usual quadratic programming method can not be directly applied to solve for c, and a nonlinear opimization technique such as the Newton=Raphson method or Fletcher=Powell method is necessary. For the specific form $w(c)$ in (3.19) or equivalently

$$(3.20) \quad w(c) = (\lambda_1 - \lambda_2)[\Sigma_{t=1}^{M} r_t(c)/T]^2 + \lambda_2 \Sigma_{t=1}^{T} r_t(c)^2/T,$$

we can proceed as follows.

i) Choose an initial value $c^0 = (c_1^0, \cdots, c_M^0)'$, and substituting

$$(3.21) \quad r_t^0(c) = [V_t(c) - V_{t-1}(c)]/V_{t-1}(c^0)$$

into (3.20) yields a quadratic form of c, which is denoted by $w^0(c)$.

ii) Maximize $w^0(c)$ by a quadratic programming technique and obtain an optimal c^1 with $c_j^1 \geq 0$. Substitute c^1 into (2.21) for c^0 and define $w^1(c)$.

iii) Repeat the procedures i) and ii) and stop when $\max_i | c_i^k - c_i^{k-1} |$ is less than a specified value. The value $c^k = (c_1^k, \cdots, c_M^k)$ is the desired portfolio unit ratio vector.

A portfolio thus constructed has an optimal performance required for the past data. We call such a portfolio an *optimal historical portfolio*. In this approach the variational processes of asset prices are not taken into consideration.

4* Mathematical structure of asset allocation

To mathematically formalize the arguments on asset allocation made in Section 1, let T denote the present time for decision making and let

$U = \{1, \cdots, M\}$ be the universe of available assets for investment which is specified in advance without observing data,

$\mathcal{Z}_T$ the information set of data available at T which necessarily includes the prices of assets in U,

$$S_t = (S_{1t}, \cdots, S_{Mt})' \quad (t = 1, \cdots, T), \text{ and}$$

$\mathcal{I}$ the set of mathematically formalizable investment stances (objective functions) which are expressions or characterizations on future behaviors of portfolios to be constructed.

Quants asset allocation problem   consists of the two stages:

(A) To choose an objective function $u \in \mathcal{X}$, and then choose a function H_u which depends on u such that $H_u(z)$ gives a portfolio $c_T = c(z_T) = (c_1(z_T), \cdots, c_M(z_T))$ for each $z_T \in \mathcal{Z}_T$:

$$H_u : \mathcal{Z}_T \quad \to \quad C = \{c = (c_1, \cdots, c_M)\}$$
$$H_u(z_T) = c(z_T), \text{ where } c_i \text{ is the } i\text{-th unit ratio.}$$

(B) To choose a trading (rebalance) rule $\{c_{T+n} \in C : n = 1, 2, \cdots\}$ as
a function of $z_{t+n} \in \mathcal{Z}_{T+n}$:

$$c_{T+n} : \mathcal{Z}_{T+n} \quad \to \quad C$$
$$c_{T+n} = c_{T+n}(z_{T+n}) = (c_1(z_{T+n}), \cdots, c_M(z_{T+n}))$$

so that the portfolio c_T formed at T is rebalanced into the
portfolio c_{T+n} at each $T+n$.

If we should know the stochastic process of asset prices S_t, we
would be able to solve the problems (A) and (B) at least approximately by a stochastic optimization technique.   However, not only
the stochastic process of S_t is unknown but also it will evolve
over time, as has been often discussed.   Under such a situation,
there will be three available quants approaches:

(1) Normative model approach.   From such a hypothesis as an efficient market hypothesis, a specific theoretical model is chosen
for analysis a priori.   A typical example is the case of option
theory in which a geometric Brownian motion (lognormal random
walk model) is assumed.

(2) Empirical model approach.   This is our approach and a statistical model which empirically performs well is to be sought
from such a set of available models as we treated in Chapters 4
and 5.

(3) Black box approach.   No model is specified and some tradig rules
are directly studied without analyzing the variational structure of the process of S_t.   Typical trading rules are filter
rule, CPR (constant proportion rule), etc. and they usually

treat univariate series.

In any approach, not all the assets in the universe are considered the objects for analysis. In particular, to construct a portfolio which performs well relative to an investment stance $u \in \mathcal{I}$, we need to select a portfolio population (subpopulation) from the universe in association with $u \in \mathcal{I}$ and analyze it possibly through a model. In the sequal, this process is discussed from the viewpoint of (A) and (B).

Problem (A): Construction of a portfolio

I   Specification of investment stance (objective function). Some examples of investment stances are listed.

i) Markowitz type return-risk objective function.

$$u(\mu_{T+n}, \ \sigma_{T+n}) \ = \ \lambda_1 \mu_{T+n}^2 + \lambda_2 \sigma_{T+n}^2,$$

where μ_{T+n} and σ_{T+n} are respectively the return and risk at $T+n$ of a portfolio which is made at T.

ii) Portfolio insurance (PI) type objective function.

$$u = \ \max(V_{T+n}(c_T), K) \qquad (n=1, \cdots, n_0),$$

where $V_{T+n}(c_T)$ is the value at $T+n$ of portfolio c_T.

iii) Index (tracking) type objective function.

$$-\alpha \ < \ x_{T+n}(c_T) - x_{0\,T+n} \ < \ \beta \quad (n=1, \cdots, n_0),$$

where $x_{T+n}(c_T)$ and $x_{0\,T+n}$ are respectively the total returns of a portfolio c_T and an index from T till $T+n_0$.

iv) Arbitrage type objective function

$$x_{T+n}(c_T) - f_{T+n} \ > \ \alpha \qquad (n=1, \cdots, n_0),$$

where f_{T+n} is the return of a futures from T till $T+n$.

In practice, we need to take the cost factors (trading cost, tax, etc.) into account.

Ⅱ Selection of a portfolio population

The future return of a portfolio directly made from the universe $U=\{1, \cdots, M\}$ is always expressed as

$$\widetilde{x}_t(d_T) = d_{1T}\widetilde{x}_{1t} + d_{2T}\widetilde{x}_{2t} + \cdots + d_{MT}\widetilde{x}_{Mt},$$

where $\widetilde{x}_{it} = S_{it}/S_{iT}$ (see (2.9)). Hence the assets which are not included in the portfolio are identified as those with $d_{iT}=0$. However, there are several reasons not to make a portfolio directly from the universe U;

i) it is costly and not practical to analyze data $\mathscr{Z}_T$ for all the assets in the universe,

ⅱ) it is not easy to make a stable portfolio directly from U which meets an investment stance and it is often the case that M is greater than sample size,

ⅲ) the inclusion of less liquid assets in the universe will cause market impact risk when bought and sold, and

ⅳ) there are institutional restrictions, e.g., on short selling, and personal restrictions on the size of initial investment fund, etc.

In particular, in the empirical model approach the reason ⅱ) is quite important for not modelling the universe directly. These reasons will require us to select a subpopulation of assets whose size is less than a certain number m. In selection of a portfolio population, we usually combine the following two methods.

a) Making a simpler analysis on the M assets maybe via a model, such assets are selected that carry certain variational features which correspond to an investment stance. Examples of such features will be high serial correlation, stochastic trend discussed in Chapter 2, etc., which makes the variations of port folios stabilized. Also selections are made based on quants predictions.

b) Such assets are selected that satisfy some given criteria (for example, PER, capitalization, etc.) which may not be directly related to an investment stance.

Mathematically such a selection procedure is viewed as a procedure choosing a population selection function F_u for a given investment stance $u \in I$;

$$F_u : \mathbb{Z}_T \rightarrow \Pi(U) = \{\pi_1, \cdots, \pi_a\} \quad (a = 2^M - 1),$$
$$F_u(z_T) = \pi(z_T) \in \Pi(U) \quad (z_T \in \mathbb{Z}_T),$$

where $\Pi(U)$ is the set of all the nonempty subsets of the universe U.

III Construction of a portfolio

Once a portfolio population $\pi_k(\in \Pi(U))$ is selected, one may proceed either

c) to modelling the price variation structure of the assets in π_k, on which an optimal portfolio is formed relative to an investment stance $u \in I$, or

d) to forming such a portfolio as equally unit-weighted portfolio, equally value-weighted portfolio, etc. without modelling.

Mathematically this procedure is viewed as a procedure choosing a portfolio making function G_u for a given $F_u(z_T) = \pi(z_T) = \{i_1, \cdots, i_m\}$;

$$G_u : \mathbb{Z}_T \times \{\pi(z_T)\} \rightarrow C = \{c = (c_1, \cdots, c_M)\},$$
$$G_u(z_T, \pi(z_T)) = (c_1(z_T), \cdots, c_M(z_T)) \in C.$$

Hence composing G_u and F_u yields

$$H_u(z_T) = G_u(z_T, F_u(z_T)) = c(z) : \mathbb{Z}_T \rightarrow C.$$

Problem (B): Selection of a trading rule

If we ignore the cost of rebalancing a portfolio, we can form an optimal portfolio $H_u(z_t) = c(z_t)$ at each t by following a

procedure in (A).   In other words, the process of { $c(z_t)$: $t = T +$
1, $T+2, \cdots$ } of choosing an optimal portfolio at each time forms a
trading rule (process).   However we cannot ignore the cost factor
and then we confront a trade-off between the optimality of a portfolio and the cost of rebalancing.   In general, a portfolio optimized at T relative to an investment stance will deviate gradually
from the investment stance as time passes.   On the other hand, if
we rebalance a portfolio at each time to obtain an optimality, the
cost will increase rapidly.   Hence we may formulate an objective
function expressing this trade-off and derive a trading rule (process), which consists of the two part; a decision on when to
rebalance and a choice of an optimal portfolio.   The optimal
portfolio $c^*(z_{T+n})$ thus obtained at a rebalance time $T + n$ is
not equal to the optimal portfolio $c(z_{T+n})$ because the rebalance
cost is taken into account in deriving $c^*(z_{T+n})$.

Once a trading rule is set up, a computer-supported system can
faithfully follow the rule without a further analysis.   From this
fact, it is sometimes said that such trading rules as a filter rule,
a portfolio insurance type trading rule etc. are prediction-free.
However, no trading rules can be prediction-free, because the rules
are predicted to perform well, maybe on the basis of simulations.

Exercises

1. State your opinion on asset allocation and compare it with the
   argument of Section 1.
2. Suppose you fix a trading rule such as a filter rule.   Do you
   think it means that it is prediction-free?
3. Show that $\widetilde{x}_t(d_T)$ in (2.10) defines the portfolio return at
   $t > T$ relative to the initial fund while $v_t(d)$ in (3.4) does
   not except for $t = T+1$.
4. Extend the Markowitz effect to a general case where there are N
   assets (see Section 3).
5. State a situation where a historical Markowitz portfolio

performs for future.

6. State some factors in selecting a portfolio population of assets from your viewpoint and discuss about the information from the time series movements. Are the factors serially correlated?

7. Do you think that a trading rule should be associated with the time series structure of returns? Or do you think that there is a profitable trading rule when x_t's are iid?

CHAPTER 7

MULTIFACTOR MODELS AND THEIR APPLICATIONS

1 Introduction

Multifactor models are popular models in portfolio analysis (see Chapter 4) and they are applied in various manners. A typical application is to create an index (tracking) portfolio, which will be discussed in Chapter 11, or a tilted portfolio, a portfolio which is responsive or nonresponsive to a specific factor such as interest rate. Multifactor models are also often associated with finance theory. In this chapter, we first review the CAPM (capital asset pricing model) in finance theory. Although the CAPM itself is derived under very strong conditions, the market model associated with the CAPM has been popular, especially in an academic world, and no doubt it has been playing an important role as a kernel (core) model leading us to alternative (modified) models. In fact, various CAPM-like models have been proposed to improve the poor empirical performance of the original CAPM market model. Some of such models are introduced in Section 2 and Rosenberg's models in Chapter 8 are also regarded as such. However, though these empirical models are modifications of the market model, they are not consistent with the CAPM theory as they stand. Those modified models are mostly multifactor models of the form

$$(1.1) \quad x_{it} = \alpha_{i0} + \alpha_{i1} f_{1t} + \cdots + \alpha_{iq} f_{qt} + \varepsilon_{it}$$

(see Chapters 4 and 5 on some statistical specifications of this model). On the other hand, in the APT (arbitrage pricing theory) treated in Section 3 it is assumed that returns follow a specific multifactor model. In fact, in a line with the efficient market hypothesis factors $\{f_{jt}\}$'s are asssumed to be mutually uncorrelated and serially uncorrelated so that no predictability is admitted from the model and the APT states that the mean return α_{i0} of each

130

asset is a linear combination of risk premiums obtained from the
exposures to each risk factor f_{jt}. In other words, in the APT the
factors f_{jt}'s are totally unpredictable (like betting on roulette)
and an association α_{ij} of the i-th return x_{it} with the j-th
risk factor f_{jt} is compensated proportionally to the size of the
risk (variance) of f_{jt}. Empirically speaking this APT model will
not well represent actual return process. Because as has been
observed in Chapter 2, returns are nonlinearly highly correlated.

Brown and Otsuki (1990) proposed what they call CAPMD (capital
asset pricing, multiple dimensional) model, which is an application
of McElroy and Burmeister (1988) approach to the APT model in Section 3. The model is adopted by the Yamaichi Securities to develop
their ISS-J (Integrative Stock System - Japan). In Section 4 the
model and system are introduced based on Brown and Otsuki (1990)
and Takano (1987). In addition, the factor analysis model applied
to an analysis of Japanese stock prices by Elton and Gruber (1988)
is introduced, upon which the Nomura Securities developed a multifactor model for portfolios. In Section 5, as an alternative
multifactor model, the MTV-Regression model is considered with an
application due to the New Japan Securities.

2 CAPM

The famous CAPM is a market equilibrium model constructed in the
framework of the Markowitz-Tobin portfolio theory.

Assumptions of the CAPM

(i) One-period model: Investors at time 0 make investment deci-
    sions for time 1 at which all the investments are closed.

(ii) Homogeneous forecasts: All the investors make a common fore-
    cast on the probability distributions of all the securities so
    that they can form the common forecasted (expected) returns and
    risks of the securities. This implies that the investors have

a common efficient frontier of portfolios.

(iii) Risk-averse investors: 1% increase in the forecasted risk must
be compensated by an increase more than 1% in the forecasted
returns.

(iv) Frictionless, complete and perfect market: No trading cost is
required, no restrictions on long and short tradings are imposed and market is perfect.

(v) Non-risk asset exists: Investors can lend and borrow as much
money as they want with a fixed interest rate r_{f1}.

Under these assumptions, the forecasted (expected) equilibrium
returns of the securities and the market portfolio satisfy the following relation

$$(2.1)\quad E(r_{i1}) = r_{f1} + \beta_{i1}[E(r_{m1}) - r_{f1}]\quad\text{with}$$
$$\beta_{i1} = \text{Cov}(r_{i1}, r_{m1})/\text{Var}(r_{m1}),$$

where r_{i1} and r_{m1} are the i-th return and the market portfolio
return at time 1 respectively. This relationship between the
returns of an individual asset and of the market portfolio is not a
causal relationship, though it is often interpreted as a causal
relation. Under (2.1), all the risk averters can attain their
utility maximization.

If the relation (2.1) holds for each period, from $t-1$ to t
the same relation can be expressed with r_{i1}, r_{m1} and β_{i1} replaced by r_{it}, r_{mt} and β_{it} respectively, from which we can define a
general relationship between r_{it} and r_{mt};

$$(2.2)\quad r_{it} = \alpha_{it} + \beta_{it} r_{mt} + \varepsilon_{it},$$

where $i = 1, \cdots, M$ and $t = 1, 2, \cdots, T$. This model is called a market
model with time-varying coefficients. It should be noted that the
model (2.2) as it stands is an identity derived from the random
variables (r_{it}, r_{mt}) which are realized simultaneously (see
Chapters 4 and 8). But for error term ε_{it}, the additional assump-

tion is usually made that

$$(2.3) \quad E(\varepsilon_{it}) = 0, \ \mathrm{Var}(\varepsilon_{it}) = \phi_{iit} \equiv \sigma^2(\varepsilon_{it}),$$
$$\mathrm{Cov}(r_{mt}, \varepsilon_{it}) = 0 \ \text{and} \ \mathrm{Cov}(\varepsilon_{it}, \varepsilon_{js}) = 0 \quad (i \neq j \ \text{or} \ t \neq s).$$

Under this assumption the role of r_{mt} is implemented as a causal variable from $t-1$ to t. In fact, (2.3) implies that

i) cross-sectional correlation is zero for each t; $\mathrm{Correl}(\varepsilon_{it}, \varepsilon_{jt}) = 0 \ (i \neq j)$,

ii) serial correlation is zero; $\mathrm{Correl}(\varepsilon_{it}, \varepsilon_{it-k}) = 0 \ (k \neq 0)$,

iii) cross serial correlation is zero; $\mathrm{Correl}(\varepsilon_{it}, \varepsilon_{jt-k}) = 0$.

Hence the expression (2.2) with (2.3) is regarded as a causal relation.

Further it also implies the relationship of variances;

$$(2.4) \quad \underset{\text{total risk}}{\sigma^2(r_{it})} = \underset{\text{systematic risk}}{\beta_{it}^2 \sigma^2(r_{mt})} + \underset{\text{unsystematic risk}}{\sigma^2(\varepsilon_{it})},$$

Since ε_{it}'s are cross-sectionally uncorrelated ($i = 1, \cdots, M$) for each t, a well diversified portfolio will reduce the variance due to the variances of ε_{it}'s by the diversification effect mentioned in Chapter 6. In fact, let d_{it-1}'s be one-period portfolio value ratios from $t-1$ to t. Then as is discussed in Chapter 6, the portfolio return at t is expressed as

$$(2.5) \quad r_{Pt} = d_{1t-1} r_{1t} + \cdots + d_{Mt-1} r_{Mt}$$

and hence from (2.2)

$$(2.6) \quad r_{Pt} = \alpha_{Pt} + \beta_{Pt} r_{mt} + \varepsilon_{Pt},$$

where $\alpha_{Pt} = \Sigma_i d_{it-1} \alpha_i$, $\beta_{Pt} = \Sigma_i d_{it-1} \beta_{it}$ and $\varepsilon_{Pt} = \Sigma_i d_{it-1} \varepsilon_{it}$. Therefore, the variance of the portfolio return (from $t-1$ to t) is

$$(2.7) \quad \sigma^2(r_{Pt}) = \beta_{Pt}^2 \sigma^2(r_{mt}) + \sigma^2(\varepsilon_{Pt}) \quad \text{with}$$

$$(2.8) \quad \sigma^2(\varepsilon_{Pt}) = d_{1t-1}^2 \sigma^2(\varepsilon_{1t}) + \cdots + d_{Mt-1}^2 \sigma^2(\varepsilon_{Mt})$$

so that a well diversified portfolio will reduce the variance σ^2 (ε_{Pt}) of portfolio error term ε_{Pt} in (2.6). By this reason σ^2 (ε_{it}) in (2.4) is called an unsystematic risk, which can be reduced by forming a diversified portfolio. However, the market risk $\sigma^2(r_{mt})$ in (2.7) is the same as the one in (2.4) and hence unless β_{it}^2's are small the risk (variance) due to $\sigma^2(r_{mt})$ cannot be reduced. By this reason, $\beta_{it}^2\sigma^2(r_{mt})$ is called a systematic risk, and β_{it}^2 is the proportion exposed to the market risk $\sigma^2(r_{mt})$. Note that this interpretation is made possible by assumption (2.3). Empirically the assumption is not supported.

So far, though time element is introduced in the market model, the model is treated as a one-period model from $t-1$ to t. If α_{it}, β_{it} and ϕ_{iit} in (2.2) and (2.3) vary with t in an intractable manner, there is no chance to estimate the model empirically. In applications the following two approaches are usually taken.

A. α_{it}, β_{it} and ϕ_{iit} are constant over time

B. α_{it} and β_{it} follow stochastic processes, but ϕ_{iit} is constant.

A Constant-coefficient market model

A constant-coefficient market model is the traditional model in the CAPM for empirical applications. The implications of this model are; (1) people's homogeneous expectations (forecasts) on future returns and risks of M assets are unchanged over time so that the CAPM relation (2.1) repeats itself each period, and (2) with the help of the assumption of (2.3) the repetitions over time are uncorrelated from one period to another period. These implications are, needless to say, very strong restrictions in reality. In particular, (2) means that the time series structures of all the individual returns $\{r_{it}\}$ are exactly the same as the time series structure of the market returns $\{r_{mt}\}$ except for the individual white noises $\{\varepsilon_{it}\}$. In other words, the systematic variation structure of each individual returns $\{r_{it}\}$ is proportional to that of the market returns $\{r_{mt}\}$ and the remaining error series $\{\varepsilon_{it}\}$ is of no time series structure (white noise) for $i=1,\cdots,M$. This will be very unrealistic without going to data. However the CAPM will basically require this structure. In fact, if we allow each $\{r_{it}\}$ to follow a different process as a stochastic process, the problem becomes a multi-period problem and an intertemporal utility maximization will be needed in the theoretical framework in which the CAPM as it stands no longer holds. Though there are many attempts to formulate the problem in a multi-period situation, it is necessary to assume that all the investors have in common the same stochastic processes of the returns of all the assets as their homogeneous forecasts, and in addition the results depend heavily on the specifications of the processes.

Now under the assumption of the constancy of the coefficients and under (2.3), the model (2.2) becomes a usual regression model:

$$(2.9) \quad r_{it} = \alpha_i + \beta_i r_{mt} + \varepsilon_{it} \quad \text{with} \quad E(\varepsilon_{it})=0,$$
$$\text{Var}(\varepsilon_{it})=\phi_{ii} \quad \text{and} \quad \text{Cov}(\varepsilon_{it}, \varepsilon_{is})=0 \quad (t \neq s).$$

Hence by the least squares (LS) method α_i, β_i and ϕ_{ii} are estimated. But as will be observed, it is often the case that the coefficient of determination is small and the t-value of β_i may not be significant. A warning is that though the relations (2.4) and (2.5) hold by dropping off the time suffix t from α_{it} and β_{it}, in order to keep the meaning of d_{it-1}'s as the portfolio value ratios from time $t-1$ to t, d_{it-1}'s must be chosen as

$$d_{it-1} = d_{iT}(S_{it-1}/S_{iT})/\sum_{j=1}^{M} d_{jT}(S_{jt-1}/S_{jT})$$

(see Chapter 6).

Empirical result on the CAPM

Following Yamada (1990), the empirical validity of the CAPM with constant coefficients is investigated. We assume

i) one-period is one-month,

ii) the market portfolio is the TOPIX (Tokyo Stock Market Price
   Index),

iii) the non-risk asset is the Nikkei Short-term Bond Index though
   its return changes over time,

iv) the period for whcih the CAPM structure is unchanged is 5 years
   (60 data).

For i) the time horizons of investments vary with risk preferences,
investment stances and types of funds. But the period in which the
CAPM is derived for an investment decision is not such a short-term
as day or week. Concerning iv), the longer the sample period for
the CAPM, the less the model will perform because the environment
including the market structure changes so that at least the coefficients of the model are likely to change. Yamada (1990) presented
the following Figure 2-1 on the transitions of betas and alphas for
Toyota and Mitsui & Co. over time. Each (α, β) is computed based
on monthly returns for past five years from the date specified in
the graphs. For example, the first beta corresponding to the date
78.1 in the graph is the least squares estimate of β for the
period 73.2~78.1 and the last one corresponding to 89.9 is based
on the data for 85.10~89.9.

Also in Figure 2-2 the t values of β's and α's are graphed based
on each five year data. From these graphs, it is easily observed
that

 (a) the betas and alphas are not stable, and

 (b) the t-values are not be significant for some periods.

B Random coefficient CAPM

   As stated in Chapter 4, the constancy of the market model over
time has been tested by Knif (1989) and many other authors, and a
common observation is that the model is not stable over time. As
alternative models some types of time-varying coefficient market

Fig. 2-1 Stability of β and α

(a) Toyota   (b) Mitsui

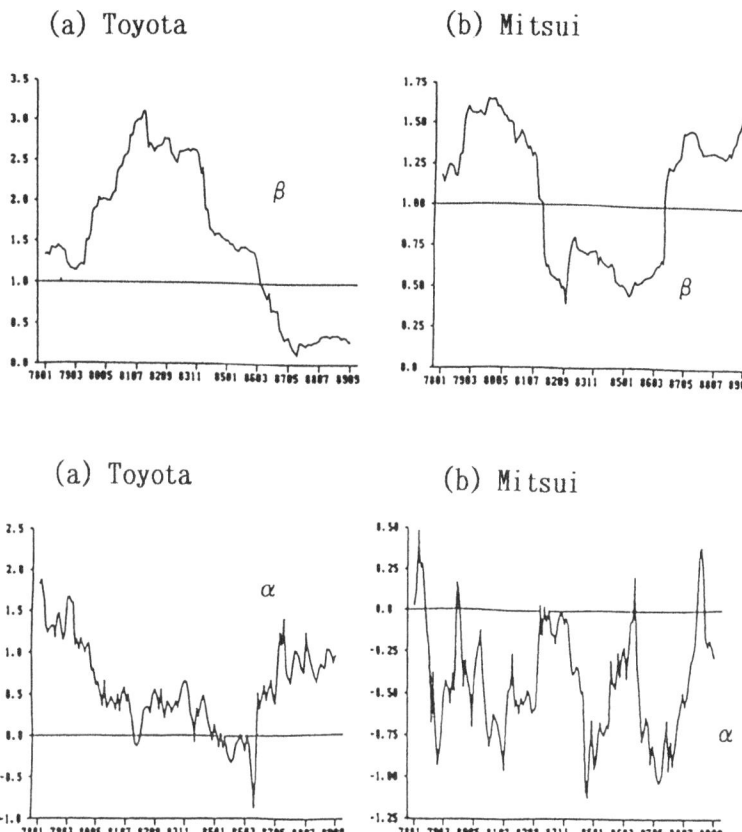

(a) Toyota   (b) Mitsui

Fig. 2-2 t values of β and α

(a) Toyota   (b) Mitsui

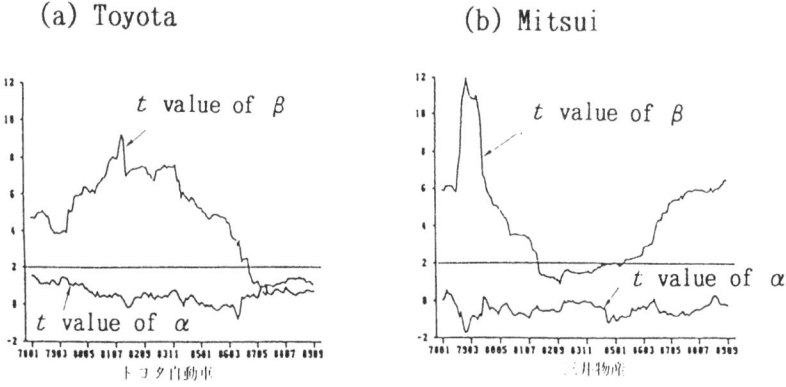

models have been considered. The following are some examples:
1) Purely random coefficient model;

$$\beta_{it} = \overline{\beta}_i + \eta_{it}, \quad \{\eta_{it}\} \text{ iid for } t=1,\cdots,T.$$

2) Stochastic parameter model;

$$\beta_{it} - \overline{\beta}_i = \theta(\beta_{it-1} - \overline{\beta}_i) + \eta_{it}, \quad \{\eta_{it}\} \text{ iid.}$$

3) Random walk model;

$$\beta_{it} = \beta_{it-1} + \eta_{it}, \quad \{\eta_{it}\} \text{ iid.}$$

In Chapter 8, a model with β_{it} depending on accounting data
and other models are introduced for asset allocation based on
Rosenberg's articles. The model 1) induces a heteroscedastic effect on the market model with constant coefficients. In the model
2) a test for the constancy $\beta_{it-1} = \overline{\beta}_i$ is given in Chapter 4.

Also a model taking into account the heteroscedasticity of ε_{it}
and some other dynamic models are
4) CAPM-ARCH;

$$h_{it} = \text{Var}(\varepsilon_{it} | \Psi_{t-1}) = \gamma_0 + \Sigma_{j=1}^q \gamma_j \varepsilon_{jt-j}^2.$$

5) Dynamic market model;

$$r_{it} = \alpha_{i0} + \alpha_{i1} r_{it-1} + \beta_{i0} r_{mt} + \beta_{i1} r_{mt-1} + \tau_i r_{mt}^2 + \varepsilon_{it}.$$

6) Dynamic error-correcting model

$$\Delta r_{it} = \alpha_{i0} + \alpha_{i1} \Delta r_{it-1} + \beta_{i1} \Delta r_{mt-1}$$
$$+ \gamma_i(r_{it-1} - \beta_{i0} r_{mt-1}) + \varepsilon_{it}.$$

In general, these models improve performance to some extent, but
not very significantly.

Recall that a market model is an identity as a statistical model
when no specification is made on error term ε_{it}. This feature
will allows us to specify

$$\alpha_{it} = \alpha_i(z_t), \quad \beta_{it} = \beta_i(z_t) \text{ and } \phi_{iit} = \phi_{ii}(z_t),$$

where z_t is a vector of fundamentals variables.   The details will be discussed in Chapter 8.

3 APT (Arbitrage pricing theory)

Another model which is paid a great attention to in finance theory is a multifactor model associated with the APT (arbitrage pricing theory), which was proposed by Roll and Ross (1980).   While the CAPM theory derives a relationship between each individual expected stock return and the expected market return defined on the efficient portfolio frontier, the APT *assumes* a multifactor model

$$(3.1)\quad x_{it} = \alpha_{i0} + \alpha_{i1} f_{1t} + \cdots + \alpha_{iq} f_{qt} + \varepsilon_{it} \quad (i=1,\cdots,N)$$

as a model for individual return generation, and derives an expression for the expected value of individual return $\alpha_{i0} = E(x_{it})$ under certain conditions with no arbitrage opportunity.

Assumptions of the APT

(i)   The model (3.1) represents the return generation from time t
   -1 to t as a *one-period model*, where the model is known to
   people with the values of coefficients α_{ij}'s.   Here factors
   f_{jt}'s are mutually uncorrelated with

   $$E(f_{jt}) = 0, \quad \text{Var}(f_{jt}) = \delta_j, \quad \text{and} \quad E(f_{jt} f_{kt}) = 0$$

   $(j \neq k)$, and they are uncorrelated with error terms ε_{it}'s,
   which satisfy

   $$E(\varepsilon_{it}) = 0, \quad \text{Var}(\varepsilon_{it}) = \sigma_{ii} \quad \text{and} \quad E(\varepsilon_{it}\varepsilon_{jt}) = \sigma_{ij}.$$

(ii)   There is no arbitrage opportunity.   An arbitrage opportunity
   is a profitable opportunity with no risk and no fund.

(iii)   The market is perfect, frictionless (no trading cost, no tax
   etc.) and complete.   Here a market is said to be complete if we
   can create any form of portfolio (short or long) with no cost.

Under these assumptions, the expected value α_{i0} of the i-th return is expressed as

(3.2) $\alpha_{i0} = E(x_{it}) \fallingdotseq \lambda_0 + \Sigma_{j=1}^q \alpha_{ij}\lambda_j$

$(i=1, 2, \cdots, N)$ when N is large. The proof is given in Appendix. The λ_j is regarded as the risk premium of the j-th factor f_{jt} which is random. In fact, let $d=(d_1, \cdots, d_N)'$ be a portfolio value ratio vector such that

(3.3) $d'1 = 0$, $d'\alpha_j = 1$ and $d'\alpha_k = 0$ $(j \neq k)$,

where $\alpha_j = (\alpha_{1j}, \cdots, \alpha_{Nj})'$ $(j, k=1, \cdots q)$. Also write (3.1) in vector notation as

(3.4) $x_t = \alpha_0 + \alpha_1 f_{1t} + \cdots + \alpha_q f_{qt} + \varepsilon_t.$

Then the portfolio return

(3.5) $r_t = d'x_t = d'\alpha_0 + f_{jt} + d'\varepsilon_t$
$= \lambda_j + f_{jt} + d'\varepsilon_t,$

because (3.2) is of the form

(3.6) $\alpha_0 = \lambda_0 1 + \Sigma_{j=1}^q \lambda_j \alpha_j.$

Therefore the portfolio tilted on f_{jt} as in (3.3) has the expected mean $E(r_t) = \lambda_j$, implying that λ_j is interpreted as the risk premium from f_{jt}.

It should be noted that the assumed APT model (3.1) with assumptions (i), (ii) and (iii) excludes the predictability for future returns from the beginning. In other words, the logical consequence (3.2) of the APT is valid only if the model with the assumptions is valid. Empirically speaking, the assumption (i) is not well supported. In many empirical studies on the APT, fundamentals factors z_{jt}'s chosen as factors f_{jt}'s are "whitened" in order for f_{jt}'s

to satisfy the assumption (i), and the residual parts after delet-
ing the systematic parts from z_{jt}'s are used (i.e., $f_{jt} \doteq (z_{jt} - E(z_{jt}))/\sigma(z_{jt})$) (see Section 4).

Estimation of an APT model and testing

The following three approaches have been customarily taken to
estimate an APT model (3.1) with (3.2), where (3.2) is often assumed to exactly hold.
1) Time-series factor analysis for $\alpha_{i,j}$ and cross-section regression for λ_j.
2) Time-series regression for $\alpha_{i,j}$ and cross-section regression
   for λ_j.
3) NLSUR (nonlinear seemingly unrelated regression) simultaneous
   inference.
The first two approaches are basically same except for the point
that factors are implicit (latent) in 1), while they are explicit
in 2). The approaches consist of the two steps:
(1) Estimation of $\alpha_{i,j}$'s. In 1), the factor analysis method is
applied to estimation of $\alpha_{i,j}$'s, where the standard assumptions
stated in Section 4 of Chapter 4 are made. In particular, it is
assumed that $f_t = (f_{1t}, \cdots, f_{qt})'$ is iid $N(0, I)$ with prespecified factor number q. On the other hand in 2), $\alpha_{i,j}$'s are estimated
by a multivariate regression model, where some observable variables
are chosen in advance for factors and the variables are transformed
so that the conditions on f_{jt}'s are satisfied.
(2) Estimation of λ_j's. For given estimated $\hat{\alpha}_{i,j}$, it is assumed
that

$$(3.7) \quad x_t = \lambda_0 1 + \hat{A}\lambda + u_t,$$

where $x_t = (x_{1t}, \cdots, x_{Nt})'$ and $\lambda = (\lambda_1, \cdots, \lambda_q)'$. Here u_t's have
$E(u_t) = 0$ and $\text{Var}(u_t) = \xi I$ and they are serially uncorrelated.
Often λ_0 is chosen to be a fixed interest rate of a safe asset in

advance, in which x_t is replaced by $x_t - \lambda_0 1$. In any case, applying a cross-sectional LS method to (3.7) with fixed t, λ is estimated as

$$(3.8) \quad \hat{\lambda}_t = (\hat{A}' \hat{A})^{-1} \hat{A}' (x_t - \lambda_0 1)$$

when λ_0 is observable. This estimate depends on t and hence it is averaged as $\bar{\lambda} = \frac{1}{T} \Sigma_{t=1}^{T} \hat{\lambda}_t$. Based on this estimate of λ, it is often considered to test the nonzero risk premium of λ_j's:

$$(3.9) \quad H: \lambda_j = 0.$$

Statistically looking at this approach, since A is estimated, $\hat{A}$ depends on x_t's and hence $\hat{A}$ and u_t are in general correlated so that the LS method may not be appropriate. This problem is known as the errors-in-variables problem.

The third approach is more direct and proposed by McElroy, et al (1985) and McElroy and Burmeister (1988). To describe it, assume that λ_0 is observable (nonobservable case is similar) and let

$$(3.10) \quad y_i = (x_{i1} - \lambda_0, \cdots, x_{iT} - \lambda_0)' \quad (i = 1, \cdots, N),$$
$$g_j = (f_{j1}, \cdots, f_{jT})' \quad (j = 1, \cdots, q), \text{ and } \quad \varepsilon_i = (\varepsilon_{i1}, \cdots, \varepsilon_{iT})'.$$

Then substituting (3.2) into (3.1) and rewriting the model as a time series regression model, we obtain for each i

$$(3.11) \quad y_i = X(\lambda) \alpha_i + \varepsilon_i \quad (i = 1, \cdots, N),$$

where

$$(3.12) \quad X(\lambda) = \lambda' \otimes 1 + G \quad \text{with}$$

$$\lambda = (\lambda_1, \cdots, \lambda_q)', \quad G = [g_1, \cdots, g_q]: T \times q \text{ and}$$
$$\alpha_i = (\alpha_{i1}, \cdots, \alpha_{iq})' \quad (i = 1, \cdots, N).$$

A joint expression of the model (3.12) is

$$
\begin{pmatrix} y_1 \\ y_2 \\ \cdot \\ \cdot \\ y_N \end{pmatrix} = \begin{pmatrix} X(\lambda) & 0 & 0 \\ 0 & X(\lambda) & \\ 0 & & X(\lambda) \end{pmatrix} \begin{pmatrix} \alpha_1 \\ \alpha_2 \\ \cdot \\ \cdot \\ \alpha_N \end{pmatrix} + \begin{pmatrix} \varepsilon_1 \\ \varepsilon_2 \\ \cdot \\ \cdot \\ \varepsilon_N \end{pmatrix}
$$

or equivalently

(3.13) $\quad y = [\, I_N \otimes X(\lambda)\,]\alpha + \varepsilon, \quad$ where

(3.14) $\quad E(\varepsilon) = 0 \quad$ and $\quad \mathrm{Var}(\varepsilon) = \Sigma \otimes I_T.$

It is noted that $\mathrm{Cov}(\varepsilon_i, \varepsilon_j) = \sigma_{ij} I$. If λ is known, this model is a (linear) SUR model proposed by Zellner (1962). Here to estimate (λ, α), an iterated nonlinear GLS (generalized LS) method is used. Let

(3.15) $\quad Q(\lambda, \alpha : \Sigma) = (y - Z(\lambda)\alpha)' [\, \Sigma \otimes I\,]^{-1} (y - Z(\lambda)\alpha),$

where $Z(\lambda) = I \otimes X(\lambda)$. First set $\lambda = 0$ and $\Sigma = I$ to get the OLSE $\hat{\alpha}^0$ and its residual $\hat{\varepsilon}^0 = Y - Z(0)\hat{\alpha}^0$. Then estimate Σ by

$$
\hat{\Sigma}^0 = (\hat{\sigma}_{ij}^0) \quad \text{with} \quad \hat{\sigma}_{ij}^0 = \frac{1}{T}\hat{\varepsilon}_i^{0\prime} \hat{\varepsilon}_j^0.
$$

Next substitute $\hat{\Sigma}^0$ into (3.15) and minimize $Q(\lambda, \alpha : \hat{\Sigma}^0)$ with respect to (λ, α) to get the estimate $(\hat{\lambda}^1, \hat{\alpha}^1)$, its residual $\hat{\varepsilon}^1 = Y - X(\hat{\lambda}^1)\hat{\alpha}^1$ and $\hat{\Sigma}^1 = (\hat{\sigma}_{ij}^1)$ with $\hat{\sigma}_{ij}^1 = \frac{1}{T}\hat{\varepsilon}_i^{1\prime} \hat{\varepsilon}_j^1$. Repeat this procedure till

$$
|(\hat{\alpha}_{ij}^k - \hat{\alpha}_{ij}^{k-1})/\hat{\alpha}_{ij}^k| < \delta \quad (i=1,\cdots, N: j=1,\cdots, q)
$$

is obtained, where δ is a prespecified small number.

   McElroy, et al. (1985) applied a result due to Gallent (1975) and showed that (λ', α') is strongly consistent and asymptotically normal with mean 0 and covariance matrix

(3.16) $\quad \Omega = \frac{1}{T}J(\lambda, \alpha)' [\, \Sigma^{-1} \otimes I_T\,] J(\lambda, \alpha)$

under some regularity conditions where

$$(3.17) \quad J(\lambda, \alpha) \equiv \left(\frac{\partial}{\partial(\lambda', \alpha')}\right)[I_N \otimes X(\lambda)]\alpha$$
$$= [A \otimes 1_T, I_N \otimes X(\lambda)]: NT \times (q+q)$$

with $A = [\alpha_1, \cdots, \alpha_q]: N \times q$. Once λ, α and Σ are estimated, the asymptotic variance Ω in (3.16) is estimated with substitution of the estimates, on which tests for the significance of λ and α are easily constructed.

McElroy and Burmeister (1988) also developed a connection bet-ween the APT model (3.1) with (3.2) and a market factor in the following manner. The error term ε_{it} in (3.1) may contain a factor associated with "market psychology" which is implicit but common to all the stocks. Hence write ε_{it} as

$$(3.18) \quad \varepsilon_{it} = \alpha_{iq+1} f_{q+1t} + \eta_{it},$$

where $\{f_{q+1t}\}$ is uncorrelated with $\{f_{jt}\}$'s ($j=1,\cdots, q$) and $\{\eta_{it}\}$ and it is serially uncorrelated. In this case, α_{iot} in (3. 2) needs to be modified as

$$(3.19) \quad \alpha_{iot} = \lambda_0 + \alpha_{i1}\lambda_1 + \cdots + \alpha_{iq+1}\lambda_{q+1}.$$

On the other hand, a one-period market return is obtained as $x_{mt} = \sum_{i=1}^N d_i x_{it}$ with value ratio ($d_1, \cdots, d_N$) and hence from (3.1) with (3.18) and (3.19)

$$(3.20) \quad x_{mt} = \lambda_0 + \lambda_m + \sum_{j=1}^q \alpha_{mj} f_{jt} + \alpha_{mq+1} f_{q+1t} + \eta_{mt},$$

where $d_i \geq 0$, $\sum d_i = 1$, $\alpha_{mj} = \sum_{i=1}^N d_i \alpha_{ij}$ ($j=1,\cdots, q+1$), $\varepsilon_{mt} = \sum_{i=1}^N d_i \eta_{it}$, and $\lambda_m = \sum_{j=1}^q \alpha_{mj}\lambda_j + \alpha_{mq+1}\lambda_{q+1}$.

When N is large and d_i's are small (well-diversified), Var(η_{mt}) is small and hence assume $\eta_{mt} = 0$. Also for the identifiability of f_{q+1t} without loss of generality assume $\alpha_{mq+1} = 1$. The market return (for one-period) is then expressed as

(3.21) $X_{mt} = \lambda_0 + \delta + \lambda_m + \Sigma_{j=1}^q \alpha_{mj} f_{jt} + f_{q+1t},$

where an additional premium δ is added (for possible market mis-pricing). Here f_{q+1t} is unobservable and serves as an error term for regression model (3.20) since f_{jt}'s are observable here. Consequently the total model consists of (3.21) and

(3.22) $X_{it} = \lambda_0 + \Sigma_{j=1}^{q+1} \alpha_{ij} \lambda_j + \Sigma_{j=1}^{q+1} \alpha_{ij} f_{jt} + \eta_{it},$

and it is estimated in the following two step procedures:
Step 1.   Estimate (3.21) by the OLS method, where λ_0 may be assumed to be known as before.   Get the estimate of $\delta + \lambda_m$.
Step 2.   Get the residual $\hat{f}_{q+1t}$ from step 1 and substitute it into (3.22) for f_{q+1t} so that all the factors become observable.   Then apply the NLSUR estimation procedure described above to get the estimates of λ and α_i's.   Then λ_m is estimated as $\hat{\lambda}_m = \Sigma_{j=1}^q \hat{\alpha}_{mj} \hat{\lambda}_j + \hat{\lambda}_{q+1}$, and hence δ is estimated as $\hat{\delta} = (\delta + \lambda_m) - \hat{\lambda}_m$.

This method is applied to the Japanese stock market by Brown and Otsuki (1991) and the model is used as the model for the Yamaichi Securities ISS-J portfolio system, which will be described in the next section.

Empirical result on the APT

McElroy and Burmeister (1988) applied the procedure described above to 70 US stocks (N=70) for the period 1972.1~1982.12. They chose T-bill rate for λ_0, the difference between returns of a 20 year government bond portfolio and a 20 year corporate bond portfolio for f_{1t}, the government bond portfolio return minus T-bill rate for f_{2t}, an unexpected deflation for f_{3t} and an unexpected growth rate in sales for f_{4t}. Upon these factors, the return (r_{mt}) on the SP500 Index are regressed;

$$\hat{r}_{mt} - \lambda_0 = .00224 - 1.330\, f_{1t} + .558\, f_{2t} + 2.286\, f_{3t} - .935\, f_{4t}$$
$$\phantom{\hat{r}_{mt} - \lambda_0 = }(.6)\quad (-3.9)\quad\quad (5.0)\quad\quad (2.0)\quad\quad (-2.3)$$
$$R^2 = .24,\quad DW = 2.13.$$

where the numbers in parentheses are t-statistics. The residual of this regression model serves as f_{5t}, an estimate of the implicit market factor. Then the APT model is estimated by the above NLSUR method and it is observed that most coefficients of f_{1t} are negatively significant, most coefficients of f_{3t} and f_{4t} are significant. The estimated risk premiums λ_i's are as follows.

	λ_1	λ_2	λ_3	λ_4	λ_5	λ_m
λ	.443%	.999%	.043%	.153%	.512%	.436%
t	4.27	4.76	1.83	2.21	3.21	1.18

Estimates are in percentage per month.

4 Two multifactor models in Japan

The APT model (with assumptions (i), (ii) and (iii)) may be used to evaluate the risk structure of an asset market associated with unexpected changes of fundamentals variables. Hamao (1988) made a parallel analysis with that of Chen, Roll and Ross (1986) by using Japanese stock data and some fundamentals data. In this section, based on Brown and Otsuki (1990) and Takano (1989), we first introduce the idea of the ISS-J (Integrative Stock System-Japan) at the Yamaichi Securities. Some of the arguments below are based on the author's inference in view of statistical theory. Secondly, the result in Elton and Gruber (1989) is introduced, whose approach is adopted by the Nomura Securities in developing a pricing model for stock returns.

Brown and Otsuki-Takano approach

Brown and Otsuki applied the nonlinear approach of McElroy and

Burmeister (1988) to the APT to capture the risk structure of
Japanese equity markets. They applied the APT model to three types
of portfolio returns rather than individual stock returns. One
type portfolio returns they considered are returns indirectly
defined through sales proportions over industries. Here they used
94 industries classified by the Yamaichi Securities, and estimated
imputed industry returns through the proportions of sales of each
firm over the industries. That is, they say "in the portfolio formation, the percentage of sales according to line of business was
collected as of the previous month and used to infer industry
returns." Though the details are not reported in the paper where
Brown and Wienstein (1983) is referred to, I suppose the industry
portfolios they call will be the portfolios obtained from the procedure described in Takano (1989), which we follow in the sequal.
Suppose each firm sells some of $M=94$ products (goods and services)
which correspond to the 94 industries. Let $w_{i,j}(t)$ be the sales
proportion of the j-th product relative to the total sales of the
i-th firm at t ($i=1,\cdots,N$; $j=1,\cdots,M$), where $w_{ik}(t)=0$ means
that the i-th firm has no sales in the k-th industry. Let $W(t)=$
($w_{i,j}(t)$) be the $N \times M$ matrix of sales proportions with $\Sigma_{j=1}^{M} w_{ij}$
$(t)=1$. It is assumed that monthly individual returns x_{it}'s are
determined from "industry returns r_{jt}'s" through the sales proportions by

$$X_{it} = w_{i1} r_{1t} + \cdots + w_{iM} r_{Mt} + \eta_{it}$$

or equivalently in vector-matrix notation

(4.1) $x_t = W(t) r_t + \eta_t$

where $r_t=(r_{1t},\cdots,r_{Mt})'$ is a vector of directly unobservable
(imputed) industry returns. But as $W(t)$ is observable (at $t-1$),
by the LS regression method r_t is estimated as

(4.2) $\hat{r}_t = (W(t)' W(t))^{-1} W(t)' x_t.$

Then it is assumed that $\hat{r}_t$ follows the APT model;

(4.3) $\hat{r}_t = \alpha_0 + A f_t + \varepsilon_t,$

which is equivalent to (3.4), where $A = [\alpha_1, \cdots, \alpha_q]$ and f_{jt}'s, ε_{kt}'s satisfy the assumption (i) in Section 3. To define factors f_{jt}'s, let

   z_{1t}: first defference in logarithms of the monthly average money supply (M1),
   z_{2t}: seasonally adjusted industrial production index (IPI),
   z_{3t}: wholesale price index (WPI),
   z_{4t}: crude oil price in \$US per barrel,
   z_{5t}: exchange rate (Yen/Dollar) defined as of month end,
   z_{6t}: one plus the month average of the overnight call rate,

and let $z_t = (z_{1t}, \cdots, z_{5t})'$. Then estimating a VAR model

(4.4) $z_t = \Pi_0 + \Pi_1 z_{t-1} + \cdots + \Pi_4 z_{t-4} + \xi_t,$

the residual $\hat{\xi}_t$ is defined to be f_t as $\hat{\xi}_t$'s are serially and cross-sectionally uncorrelated in their case.

   In the case of Brown and Otsuki (1990) the sample period for the APT is from June 1980 through August 1988, though the VAR model is estimated based on data from January 1978 to August 1988. Applying the McElroy and Burmeister procedure described in Section 3, they obtained the following result for the risk premiums λ_i's ($i = 1, \cdots, 6$) and λ_7 for residual market factor

	α_0	λ_1	λ_2	λ_3	λ_4	λ_5	λ_6	λ_7
OLS	1.005	.00038	.00211	.00332	-.02294	.00021	.01499	.81683
	(4.1)	(0.2)	(0.8)	(3.4)	(-2.8)	(0.0)	(2.0)	(2.3)
SUR	1.284	-.00002	.00119	.00289	-.02054	.00040	.01557	.50526
	(11.4)	(-0.0)	(2.6)	(10.5)	(-9.9)	(0.2)	(7.8)	(3.7)

The numbers in parentheses are t-values. They also reported (1) industries with most significant negative exposures, (2) industries with most significant positive exposures to macro residual factors f_{jt}'s in terms of the t-values of SUR factor coefficients α_{jk} of $\hat{r}_{jt}$ in (4.2);

Macro residual	(1)	(2)
Money Supply	Textile (-1.47)	Securities (3.38)
IPI	Real Estate (-2.34)	Refined Oil (1.87)
WPI	Crude Oil (-2.31)	Construction (1.66)
Crude Oil	Trust Banks (-3.24)	Batteries (2.07)
Exchange Rate	Consumer Credit (-3.19)	Precision Machinery (3.10)
Call Rate	Processed Foods (-3.56)	Electric Power (2.09)
Market Factor		Securities (11.87)

t values in parentheses.

Takano (1989) described some procedures for forming portfolios through the model thus obtained. For example, substituting (4.3) into (4.1) yields

$$x_t = W(t) \alpha_0 + W(t) A f_t + W(t) \varepsilon_t + \eta_t.$$

Hence the responses of x_{it} to f_{jt} is obtained as the (i, j) element of $W(t) A$ with A estimated by the NLSUR method, and so such a portfolio which does not respond to unexpected changes of oil prices can be formed. In fact, as $d' x_t$ is a portfolio return from t to $t+1$ as discussed in Chapter 6 where $d = (d_1, \cdots, d_N)'$ is a vector of portfolio value ratios, imposing $b_4 = 0$ in $d' W(t)$ $A = b = (b_1, \cdots, b_7)'$ on d will yield such a portfolio where an objective function can be optimized. Conversely a tilted portfolio which is very responsive to a specific unexpected factor can be constructed. It is remarked that the model is a risk model, meaning that it excludes the predictability. But if one could predict some future changes of a specific variable unpredicted (unexpected) by the public, the model could be used to exploit returns due to the unexpected change by making an optimal portfolio tilted on the variable.

Factor analysis model

Elton and Gruber (1988) viewed the factor analysis model described in Chapter 4 as a return generating model and applied it to the analysis of returns of 393 stocks contained in the NRI (Nomura Research Institute) 400 stock (value weighted) index. The 393 stocks were broken into 4 groups of 98, 98, 98 and 99 stocks because of the limitations of the maximum likelihood factor analysis program and because of the advantages of testing factor structures across different samples. Their sample is 179 monthly return data from May 1971 to March 1986.

They first carefully identify the number of common factors generating individual returns with 4 in the following three methods. The first one is Schwartz's Bayesian criterion (see Chapter 4 for the criterion in case of VAR model). The criterion leads to a three or four factor model for each group. Secondly they applied canonical correlation analysis, which will be discussed in Chapter 11 in a different context. Thirdly out of the 393 stocks in the four goups 20 portfolios based on size are formed every month and one month portfolio returns are computed. Then the portfolio returns are regressed against the two, three, four, five and six factor solution from each of four groups. The average R^2 (determination coefficient) for each of the four groups over the 20 groups are obtained as follows.

factors	2	3	4	5	6
Group 1	0.686	0.713	0.771	0.772	0.772
Group 2	0.663	0.700	0.763	0.777	0.779
Group 3	0.673	0.681	0.764	0.768	0.769
Group 4	0.678	0.731	0.763	0.770	0.775

From these results they conclude that a four factor model is most promissing for all the groups.

Next they observe that the four factors extracted from the first group are better representative among the sets of the four factors of the four groups though four factors from each group are more or less similar.

Thirdly they compare the explanatory powers of the four factor model with those of a market model applied to each of the 20 portfolios, where the NRI index is adopted as a market index. An observation from this comparison is of course that the four factor model dominates the market model overall in terms of adjusted R^2 though the first three portfolios are better explained by the market model because of the size effect of portfolios. A second observation is that the second, third and fourth factors take care of differences in the sizes of the portfolios and relatively smaller firms have lower betas in the market model, which is contrary to the US case, though the firms adopted in the NRI index are all fairly large. Also return is strongly related to size and the smaller the size is, the larger the one-month return. The stability of the coefficients (sensitivities) to factors are also checked.

Finally they compare the out-of-sample performances of the four factor model with those of the market model by making index tracking portfolios. They choose Nikkei 225 over the five years from January 1, 1981 to December 31, 1986 to match. For the four factor model a factor analysis is run at the beginning of each quarter using the prior 11 years of return data to extract the four factors and the 393 individual return series and Nikkei 225 returns are regressed on the four factors over the same period. Similarly at the beginning of each quarter market models are obtained via regression for the 393 stocks and the Nikkei 225 over the previous five years where the Tokyo Stock Exchange Index is adopted as a market index. Then optimal index (tracking) portfolios are obtained for each model as follows. Let

$$x_{it} = \alpha_{i0} + \alpha_{i1} f_{1t} + \cdots + \alpha_{i4} f_{4t} + \varepsilon_{it}$$

be the four factor model where $i = 0$ corresponds to the Nikkei 225. Then an optimal index portfolio is a portfolio $d = (d_1, \cdots, d_N)'$ with return $d' x_t$ which minimizes $\sum_{t=1}^{T} (\sum_{i=1}^{N} d_i \varepsilon_{it})^2$ subject to

$$\sum_{i=1}^{N} d_i \alpha_{ij} = \alpha_{0j} \quad (j=0,1,\cdots,4) \quad \text{and} \quad 0 \leq d_i \leq D.$$

Similarly an optimal index portfolio is obtained based on the market model. Then they show that the out-of-sample performances of the index portfolio based on the four factor model is much better than those on the market model though they did not give the absolute tracking errors of the index portfolios. In their approach, the portfolios seem to be rebalanced quarterly.

A further consideration into the construction of index tracking portfolios is given in Chapter 11. Tsuda (1991) made a parallel analysis by using industry indices for Japanese stocks.

5 Regression-MTV model

As has been stated in Chapter 6, the most important part in quants portfolio system is the part of modelling price variations and when the model is not satisfactory, a portfolio constructed on it will not be expected to perform well. In this section, we propose an alternative multifactor model, which will be a more feasible model to make portfolios.

As has been stated, a multifactor model of the form

$$(5.1) \quad x_{it} = \alpha_{i0} + \alpha_{i1} z_{1t} + \cdots + \alpha_{iq} z_{qt} + \varepsilon_{it} \quad (i=1,\cdots, p)$$

is basically treated in the two ways;

a) factors $z_{1t}, \cdots, z_{qt}$ are implicit and only price or return data
   is analyzed through the MTV model or factor analysis model, etc.,

b) factors $z_{1t}, \cdots, z_{qt}$ are explicit and fundamentals and accounting data for factors are used via regression.

A model we propose here is of a combined structure.

To explain our model, it is noted that

(1) in a practical application of the model (5.1) it is important to select a suitable sample period and proper transformation of variables for the approach b),

(2) x_{it}'s, which represent returns or prices or log-prices etc., follow time series processes and hence it is important to specify ε_{it}'s properly,

(3) when x_{it}'s are individual asset returns, the explanatory power of the model, whether it may be based on approach a) or approach b), is not large (low coefficient of determination),

(4) to forecast x_{it}, it is necessary to forecast z_{it}'s and

(5) in approach b) there may be correlations among the explanatory variables such as exchange rates and interest rates (possibility of multicollinearity).

A method to treat the problems associated with (2)∼(5) is as follows. First apply MTV models to data ($x_{1t}, \cdots, x_{pt}$) and data ($z_{1t}, \cdots, z_{qt}$) separately where $t=1, \cdots, T$, so that we have

$$(5.2) \quad x_{it} = \mu_{xi} + \beta_{i1} f_{1t} + \cdots + \beta_{ik} f_{kt} + \eta_{it},$$

$$(5.3) \quad z_{jt} = \mu_{zj} + \gamma_{j1} g_{1t} + \cdots + \gamma_{jr} g_{rt} + \nu_{jt}$$

($i=1, \cdots, p$, $j=1, \cdots, q$) and then regress each factor f_{at} on factors $g_{1t}, \cdots, g_{rt}$ as

$$(5.4) \quad f_{at} = \tau_{a1} g_{1t} + \cdots + \tau_{ar} g_{rt} + \psi_{at}$$

($a=1, \cdots, k$). Or more directly we can regress each x_{it} on the factors $g_{1t}, \cdots, g_{rt}$ as

$$(5.5) \quad x_{it} = \theta_{i0} + \theta_{i1} g_{1t} + \cdots + \theta_{ir} g_{rt} + \lambda_{it},$$

which is known as principal component regression. Since g_{bt}'s are supposed to follow time series models such as AR, ARCH, etc. and hence we can forecast x_{it}'s by modelling the time series structure of g_{bt}'s. Also this approach reveals the time series structures behind variables ($x_{1t}, \cdots, x_{pt}$) and variables ($z_{1t}, \cdots, z_{qt}$) and their relations through which we will be able to make a so-called tilted portfolio under a certain investment stance. In other words, a portfolio whose variations are more sensitive to a specific (fundamental) variable such as interest rate can be created. However, when a regression model in (5.1) does not have a large explanatory power, the model in (5.4) does not have a large explanatory power either, in which case a portfolio constructed through the model will not have a stable performance.

To overcome this problem, we need to model the error terms ε_{it}'s in (5.1) because we are not sure if the multivariate time series variation structures of x_{it}'s are fully captured by those of z_{it}'s even though they are captured to some extent. Here we propose to model ε_{it}'s by an MTV model as

$$(5.6) \qquad \varepsilon_{it} = \delta_{i1} h_{1t} + \cdots + \delta_{is} h_{st} + \phi_{it} \qquad (i=1, \cdots, p)$$

so that we have a multivariate regression-MTV model

$$(5.7) \qquad x_{it} = \alpha_{i0} + \alpha_{i1} z_{1t} + \cdots + \alpha_{iq} z_{qt}$$
$$+ \delta_{i1} h_{1t} + \cdots + \delta_{is} h_{st} + \phi_{it} \qquad (i=1, \cdots, p).$$

Recall that p is not small in applications so that each implicit time series variance component h_{kt} in error terms ε_{it}'s will be likely to form a common factor exhibiting some systematic movement. The multivariate regression-MTV model becomes a multivariate regression model when $\delta_{i1} = \cdots = \delta_{is} = 0$ and it becomes an MTV model when $\alpha_{i1} = \cdots = \alpha_{iq} = 0$. This model has a possibility of involving the problems (4) and (5), for which we may use

$$(5.8) \qquad x_{it} = \alpha_{i0}' + \alpha_{i1}' g_{1t} + \cdots + \alpha_{ir}' g_{rt}$$
$$+ \delta_{i1} h_{1t} + \cdots + \delta_{is} h_{st} + \phi_{it}$$

where g_{jt}'s are the variance components extracted from (5.3). An estimation procedure of (5.7) is given later.

Application

The New Japan Securities adopted the regression MTV model (5.8) to obtain a multifactor model for constructing tilted portfolios and index portfolios. As macroeconomic fundamentals variables z_{jt}'s, they used

(1) current balance of trade (z_{1t}), (2) crude oil price (z_{2t}), (3) industrial production index (z_{3t}), (4) inflation rate (z_{4t}), (5) money supply (z_{5t}), (6) T-bond rate (z_{6t}), (7) call rate (z_{7t}), (8) Yen/Dollar exchange rate (z_{8t}), (9) US inflation rate (z_{9t}), (10) US T-bill rate (z_{10t}).

Sugano, et.al. (1990) reported the following results. For x_{it}'s, prices of 224 stocks in the Nikkei 225 (excluding NTT) are analyzed with monthly data over the following 8 periods of 5 years;

I 83.4~88.3, II 83.7~88.6, III 83.10~88.9, IV 84.1~88.12
V 84.4~89.3, VI 84.7~89.6, VII 84.10~89.9, VIII 85.1~89.12.

First from the 10 variables z_{jt}'s MTV time series variance components $\{g_{kt}\}$'s are extracted via (5.3), where the cumulative rate of contribution is required to be more than 95%. Next, 224 stock prices are separately regressed on these components for each of the 8 periods to get the residuals ε_{it}'s, to which an MTV model is applied to extract residual MTV time series variance components $\{h_{mt}\}$'s as in (5.6), where 95% is the level of explanatory power. These residual components are identified as residual industry factors as follows;

1) Nikkei 36 industry indices are regressed on g_{kt}'s to extract residuals of industry indices,

2) MTV components, say $\{\widetilde{h}_{mt}\}$'s, are again extracted from these residuals for each period,

3) h_{mt}'s are regressed on $\widetilde{h}_{mt}$'s for each period to get very high explanatory power.

Based on this identification, they use the regression model

(5.9) $x_{it} = \alpha_{i0} + \alpha_{i1}g_{1t} + \cdots + \alpha_{ir}g_{rt}$
$$+ \delta_{i1}\widetilde{h}_{1t} + \cdots + \delta_{is}\widetilde{h}_{st} + \phi_{it}$$

for individual stock prices ($i=1,\cdots,224$) rather than the model
(5.8) which uses implicit factors h_{mt}'s. In this model, $\widetilde{h}_{mt}$'s
are interpreted as residual common industry factors. The numbers
r and s of fundamental common components $\{g_{jt}\}$'s and of residual
common industry factors $\{\widetilde{h}_{mt}\}$'s extracted for each period are as
follows.

period	I	II	III	IV	V	VI	VII	VIII
r	3	3	3	3	3	3	4	4
s	9	8	8	7	7	6	5	4
total	12	11	11	10	10	9	9	8

It is observed from this table that the number of the residual
industry factors $\{h_{mt}\}$'s is monotonically decreasing over the 8
periods, which will imply a more syncronized movement among the
industry residuals in 2) in recent years. Also the number r of
the fundamental common components $\{g_{jt}\}$'s indicates a high correla
tion structure among the 10 fundamental variables z_{jt}'s.

The regression results of 224 stocks over each period via (5.9)
are summarized as follows.

	I	II	III	IV	V	VI	VII	VIII
$\bar{R}^2$	.939	.942	.947	.946	.949	.952	.952	.948
SD	.070	.060	.056	.058	.052	.046	.046	.076

Here $\bar{R}^2$ is the average of adjusted R_i^2's (determination coefficients) of 224 regression equations over each period and SD is the
standard deviation $[\Sigma(R_i^2 - \bar{R}^2)^2/224]^{1/2}$. This result shows that

the components g_{jt}'s and $\widetilde{h}_{mt}$'s are jointly of high explanatory power over stock prices. The stability of the regression coefficients in (5.9) are carefully checked in Sugano et.al.(1990). It is remarked that the TOPIX (Tokyo Stock Price Index) is perfectly explained by the components extracted via (5.9) for each period. For example, for period Ⅷ,

$$TOPIX_t = 1816 - 189.5\,g_{1t} - 156.7\,g_{2t} - 265.6\,g_{3t} + 33.8\,g_{4t}$$
$$+ 18.5\,\widetilde{h}_{1t} + 11.3\,\widetilde{h}_{2t} - 12.2\,\widetilde{h}_{4t} + \varepsilon_{it}$$
$$R^2 = 1.00, \quad DW = 1.42,$$

where variables are selected based on t-values.

Estimation of a multivariate regression-MTV model

A. Let $y_i = (x_{i1}, \cdots, x_{iT})'$ $(i = 1, \cdots, p)$, $y = (y_1', \cdots, y_p')'$: $Tp \times 1$, $z_j = (z_{j1}, \cdots, z_{jT})'$ and $Z = (z_1, \cdots, z_q) : T \times q$. Then the model (5.1) is expressed as

$$(5.10) \quad y = [I \otimes Z]\alpha + \varepsilon,$$

where $\alpha = (\alpha_1', \cdots, \alpha_p')'$ with $\alpha_i = (\alpha_{i1}, \cdots, \alpha_{iq})'$, and $\varepsilon = (\varepsilon_1', \cdots, \varepsilon_p')'$ with $\varepsilon_i = (\varepsilon_{i1}, \cdots, \varepsilon_{iT})'$. Hence the LS estimate of α and the LS residual are given by

$$(5.11) \quad \hat{\alpha}_0 = (I \otimes (Z'Z)^{-1})(I \otimes Z')\,y \quad \text{and} \quad e = y - (I \otimes Z)\hat{\alpha}_0$$

respectively, where $e = (e_1', \cdots, e_p')'$ with $e_i = (e_{i1}, \cdots, e_{iT})'$. Apply an MTV model to $\{e_{it}\}$ with covariance matrix

$$\Sigma_0 = (\hat{\sigma}_{0ij}) \quad \text{where} \quad \hat{\sigma}_{0ij} = \frac{1}{T}\Sigma_{t=1}^{T} e_{it}\,e_{jt}$$

to obtain $e_{it} = \hat{\delta}_{0i1} h_{01t} + \cdots + \hat{\delta}_{0is} h_{0st} + \phi_{0it}$. Then the autocovariance matrix of $\{h_{0jt}\}$ is estimated by

$$\widehat{\Gamma}_{0j} = (\hat{\gamma}_{0juv}) \quad \text{with} \quad \hat{\gamma}_{0juv} = \frac{1}{T}\Sigma_{t=u-v+1}^{T} h_{0jt}\,h_{0jt-u+v},$$

where $u > v$. Therefore the covariance matrix of ε is estimated by

$$\hat{\Omega}_0 = \sum_{j=1}^{s} \hat{\delta}_{o\,j} \hat{\delta}_{o\,j}{}' \otimes \hat{\Gamma}_{o\,j} + D \quad \text{where}$$

$$\hat{\delta}_{o\,j} = (\hat{\delta}_{o1\,j}, \cdots, \hat{\delta}_{op\,j})' \quad \text{and}$$

$$D = \text{diag}\{\hat{\pi}_1, \cdots, \hat{\pi}_p\} \quad \text{with} \quad \hat{\pi}_i = \frac{1}{T} \sum_{t=1}^{T} \phi_{0\,i\,t}^2.$$

Of course, the number s of common components must be chosen in advance based on rate of cumulative contribution, etc.

B. Next the coefficient vector α is re-estimated by the GLS method as

$$\hat{\alpha}_1 = [(I \otimes Z)' \hat{\Omega}_0^{-1} (I \otimes Z)]^{-1} (I \otimes Z)' \hat{\Omega}_0^{-1} y$$

and the GLS residual vector is obtained as $\hat{\varepsilon} = y - (I \otimes Z) \hat{\alpha}_1$. Therefore we can obtain an MTV model for $\{\hat{\varepsilon}_{i\,t}\}$;

$$\hat{\varepsilon}_{i\,t} = \hat{\delta}_{i1} h_{1\,t} + \cdots + \hat{\delta}_{is} h_{s\,t} + \phi_{i\,t}.$$

The procedure may be repeated until a stable result is obtained.

Exercises

1. Discuss about the implications of the assumptions (3.2a) and (3.2b) in association with the CAPM. Note that $r_{i\,t}$'s and $r_{m\,t}$ are simultaneously realized and that $r_{m\,t}$ is a function of $r_{i\,t}$'s.
2. Show (2.7).
3. Discuss about the impications of the APT assumption (i) in association with the efficient market hypothesis in Chapter 1.
4. Give some reasons why beta's in the CAPM are not stable empirically. In what way can random coefficient CAPM models improve the original CAPM model?
5. How do you associate the APT assumption (i) with the empirical fact in Chapters 2 and 3 that returns are in general of a nonlinear structure?

CHAPTER 8

B. ROSENBERG MODELS AND THEIR APPLICATIONS

1 Market model portfolio

The CAPM discussed in Chapter 7 has been providing associated CAPM-like models though most of them are not consistent with the assumptions and framework of the CAPM. Such a model is a time-varying coefficient market model described in Chapter 7. In this chapter, from a viewpoint of portfolio quants, we shall overview some time-varying coefficient models proposed by Rosenberg and related models. We first review some basic concepts in this chapter.

(1) Let (x, y) be a pair of random variables with mean and covariance matrix;

$$(1.1) \quad E\begin{pmatrix} x \\ y \end{pmatrix} = \begin{pmatrix} \mu_x \\ \mu_y \end{pmatrix}, \quad \text{Cov}\left(\begin{pmatrix} x \\ y \end{pmatrix}\right) = \begin{pmatrix} \sigma_{xx} & \sigma_{xy} \\ \sigma_{yx} & \sigma_{yy} \end{pmatrix},$$

and define

$$(1.2) \quad \begin{aligned} \beta &= \text{Cov}(y, x)/\text{Var}(x) = \sigma_{xy}/\sigma_{xx}, \\ \alpha &= \mu_y - \beta\mu_x, \quad \text{and} \quad \varepsilon = y - \alpha - \beta x. \end{aligned}$$

Then it is easy to obtain the identity;

$$(1.3) \quad \begin{aligned} y &= \alpha + \beta x + \varepsilon \quad \text{with} \quad E(\varepsilon) = 0, \\ \sigma_y^2 &= \beta^2 \sigma_x^2 + \sigma_\varepsilon^2, \quad \text{and} \quad \text{Cov}(x, \varepsilon) = 0, \end{aligned}$$

where $\sigma_y^2 = \sigma_{yy}$, $\sigma_x^2 = \sigma_{xx}$ and $\sigma_\varepsilon^2 = \sigma_{yy} - \sigma_{yx}\sigma_{xx}^{-1}\sigma_{xy} = \text{Var}(\varepsilon)$. This implies that for any pair (x, y), a market model of the form (1.3) is obtained, and it does not express a causal relationship of x and y. Further, in most situations x and y are simultaneously realized with correlation $\rho_{xy} = \sigma_{xy}/\sigma_x\sigma_y$.

(2) Suppose that random variables $z_1, \cdots, z_n$ are mutually uncorrelated with variance $\sigma_i^2 = \text{Var}(z_i)$. Then for any constants

159

$a_1, \cdots, a_n,$

$$\text{Var}(\Sigma_{i=1}^{n} a_i z_i) = \Sigma_{i=1}^{n} a_i^2 \text{Var}(z_i) = \Sigma_{i=1}^{n} a_i^2 \sigma_i^2.$$

If $|\sigma_i^2| \leqq K$ and $a_i \approx 1/n$, then

(1.4) $\text{Var}(\Sigma_{i=1}^{n} a_i z_i) \leqq K/n \to 0$ as $n \to \infty,$

which was called the diversification effect in Chapter 6, as opposed to the Markowitz effect.

Now let r_{it} be the return of the i-th asset from $t-1$ to t and let

(1.5) $E(r_{it}) = \mu_{it},$ $\text{Cov}(r_{it}, r_{jt}) = \begin{array}{l} \sigma_{it}^2 \quad (i=j) \\ \sigma_{ijt} \quad (i \neq j), \end{array}$

with $\sigma_{ijt} = \sigma_{it} \sigma_{jt} \rho_{ijt}$, where ρ_{ijt} is the correlation between r_{it} and r_{jt}, $\sigma_{it}^2 = \sigma_{iit}$ and $i, j = 1, \cdots, N$. Also let

(1.6) $r_{mt} = \Sigma_{i=1}^{N} d_{mi} r_{it}$ and $r_{at} = \Sigma_{i=1}^{N} d_{ai} r_{it}$

denote a one-period market portfolio return and another one-period portfolio return respectively. Then by (1), the following identity for (r_{it}, r_{mt}) holds:

(1.7) $r_{it} = \alpha_{it} + \beta_{it} r_{mt} + \varepsilon_{it}$ $(i = 1, \cdots, N),$

$\beta_{it} = \Sigma_{j=1}^{N} d_{mj} \sigma_{ijt} / \Sigma_{j=1}^{N} \Sigma_{k=1}^{N} d_{mj} d_{mk} \sigma_{ikt}$

$\alpha_{it} = \mu_{it} - \beta_{it} \mu_{mt},$ $\text{Cov}(\varepsilon_{it}, r_{mt}) = 0,$

$\sigma_{it}^2 = \beta_{it}^2 \sigma_{mt}^2 + \phi_{iit}$ with $\phi_{iit} = \text{Var}(\varepsilon_{it}),$

and also for (r_{at}, r_{mt}) we obtain the identity;

(1.8) $r_{at} = \alpha_{at} + \beta_{at} r_{mt} + \varepsilon_{at},$

$\beta_{at} = \text{Cov}(r_{at}, r_{mt}) / \text{Var}(r_{mt}) = \Sigma_{i=1}^{N} d_{ai} \beta_{it},$

$\alpha_{at} = \Sigma_{i=1}^{N} d_{ai} \alpha_{it},$ $\text{Cov}(\varepsilon_{at}, r_{mt}) = 0,$

$\sigma_{at}^2 = \beta_{at}^2 \sigma_{mt}^2 + \phi_{aat}$ with $\phi_{aat} = \text{Var}(\varepsilon_{at}),$

$\varepsilon_{at} = \Sigma_{i=1}^{N} d_{ai} \varepsilon_{it}.$

In (1.7) $\beta_{it}^2 \sigma_{mt}^2$ is often referred to as the "systematic" risk of the i-th asset because the market risk σ_{mt}^2 cannot be reduced by making a portfolio and the exposure β_{it}^2 to the market risk is regarded as "structurally" determined by (1.7). While ϕ_{iit} is often called the "unsystematic" risk, meaning that it can be reduced by making a diversified portfolio. The first interpretation concerning $\beta_{it}^2 \sigma_{mt}^2$ will be true once a specific market return r_{mt} is chosen to obtain the identity (1.7). However, the systematic part $\beta_{it}^2 \sigma_{mt}^2$ is usually very small compared to the variance σ_{it}^2 of the i-th return and sometimes negligible in applications. In fact, the determination coefficient of regression (1.7) which estimates $\beta_{it}^2 \sigma_{mt}^2 / \sigma_{it}^2 \equiv \rho_{imt}^2$ is usually very small. Hence most information concerning the variation of r_{it} lies in the remainder term ε_{it} of the identity and $\{\varepsilon_{it}\}$ will not be a white noise. This will imply that ε_{it}'s may not be uncorrelated, so that the second interpretation on the unsystematic part will not hold as it stands. To discuss this point further, let

$$(1.9) \quad \begin{aligned} \boldsymbol{d}_m &= (d_{m1}, \cdots, d_{mN})', \quad \boldsymbol{d}_a = (d_{a1}, \cdots, d_{aN})', \\ \boldsymbol{r}_t &= (r_{1t}, \cdots, r_{Nt})', \quad \boldsymbol{\beta}_t = (\beta_{1t}, \cdots, \beta_{Nt})', \\ \boldsymbol{\alpha}_t &= (\alpha_{1t}, \cdots, \alpha_{Nt})', \quad \boldsymbol{\varepsilon}_t = (\varepsilon_{1t}, \cdots, \varepsilon_{Nt})', \\ \mathrm{Cov}(\boldsymbol{r}_t) &= \Sigma_t = (\sigma_{ijt}), \quad \mathrm{Cov}(\boldsymbol{\varepsilon}_t) = \Phi_t = (\phi_{ijt}). \end{aligned}$$

Then (1.7) is expressed as a one-factor model;

$$(1.10) \quad \begin{aligned} \boldsymbol{r}_t &= \boldsymbol{\alpha}_t + \boldsymbol{\beta}_t \, r_{mt} + \boldsymbol{\varepsilon}_t \quad \text{with} \quad r_{mt} = \boldsymbol{d}_m' \, \boldsymbol{r}_t, \\ \Sigma_t &= \boldsymbol{\beta}_t \sigma_{mt}^2 \boldsymbol{\beta}_t' + \Phi_t \quad \text{with} \quad \sigma_{mt}^2 = \boldsymbol{d}_m' \, \Sigma_t \, \boldsymbol{d}_m, \end{aligned}$$

and with $r_{at} = \boldsymbol{d}_a' \, \boldsymbol{r}_t$ (1.8) is expressed as

$$(1.11) \quad \begin{aligned} \boldsymbol{d}_a' \, \boldsymbol{r}_t &= \boldsymbol{d}_a' \, \boldsymbol{\alpha}_t + \boldsymbol{d}_a' \, \boldsymbol{\beta}_t \, r_{mt} + \boldsymbol{d}_a' \, \boldsymbol{\varepsilon}_t, \\ \boldsymbol{d}_a' \, \Sigma_t \, \boldsymbol{d}_a &= \mathrm{Var}(\boldsymbol{d}_a' \, \boldsymbol{r}_t) = (\boldsymbol{d}_a' \, \boldsymbol{\beta}_t)^2 \sigma_{mt}^2 + \boldsymbol{d}_a' \, \Phi_t \, \boldsymbol{d}_a. \end{aligned}$$

In this notation, when $\boldsymbol{\varepsilon}_t$ is referred to as an unsystematic part, Φ_t is at least implicitly assumed to be diagonal so that ε_{it} and

ε_{jt} are uncorrelated.   This presumption is often regarded as a
natural conclusion of the CAPM theory though the theory never
states it.   Even if we use the identity (1.10), there will be many
other factors which are commonly related to r_{it}'s and enables us
to model the "remainder term" ε_t.   This will lead us to a multifactor model, in which the covariance structure of Φ_t is specified.
We emphasize the identity feature of the model because r_{it}'s and
r_{mt} are simultaneously realized and because Σ_t in (1.10) is simply a reparametrization of Σ_t unless we add some assumptions such
as the diagonality of Φ_t.   If Φ_t is not diagonal, $\phi_{aat} = d_a' \Sigma_t$
d_a is not necessarily made smaller by the diversification effect
but it may be made smaller by the Markowitz effect (Section 2 of
Chapter 6).   In addition, there may be a trade-off in making $d_a' \Phi_t$
d_a and $(\beta_t' d_a)^2$ smaller.

2 Time-varying coefficient models

Rosenberg (1973a) classified some time-varying coefficient
models according to the rationales behind the models, which may be
stated as follows.
1) Coefficients follow stochastic processes for some reasons.
2) In a cross-sectional and time-series model representing individual befavior, the coefficient of each individual at each time
   t may be cross-sectionally regarded as a random sample from a
   time-invariant distribution.
3) Specification errors of a model may be treated by randomizing
   coefficients.
4) Because of an aggregation problem in the model, coefficients are
   regarded as random.
5) (Lucas) Policy makers and individual agents interact each other
   so that parameters change (game-theoretic uncertainty).
Let

$$(2.1) \quad y_{it} = x_{it}' \beta_{it} + \varepsilon_{it} \quad\quad (i=1,\cdots, N: \ t=1,\cdots, T)$$

be a cross-sectional and time-series regression model, where x_{it}: $k \times 1$. Rosenberg (1973a) classifies the time-varying coefficient feature of this model as follows.

(1) Systematic variation model for β_{it};

$$(2.2) \quad \beta_{it} = f(Z_{it}:\lambda) \simeq Z_{it}\gamma + \omega_{it}$$

so that

$$(2.3) \quad y_{it} = x_{it}' Z_{it}\gamma + x_{it}'\omega_{it} + \varepsilon_{it},$$

where Z_{it}: $k \times \ell$ is a fixed matrix and ω_{it} is independent of ε_{it}. Here ω_{it} is a random factor so that it is an error term of the approximation in (2.2). This implies that the variance of y_{it}

$$(2.4) \quad \mathrm{Var}(y_{it}) = x_{it}'\mathrm{Var}(\omega_{it})x_{it} + \mathrm{Var}(\varepsilon_{it})$$

is heteroscedastic over time even if $\mathrm{Var}(\varepsilon_{it})$ is constant.

(2) Stochastic variation model for β_{it};

$$\beta_{it} = \Gamma\beta_{it-1} + \eta_{it}.$$

This is a model treated in Chapters 4 and 7.

(3) Randomly dispersed parameter model for β_{it};

$$\beta_{it} = \mu + \eta_i \quad \eta_i \text{ iid } (i = 1, \cdots, N).$$

In the literature Rosenberg (1973b) treated a cross-section and time series model, while Swamy (1970) treated a model in which individual parameters are randomly dispersed across the population but constant over time as in (3). A model with $\beta_{it} = \gamma_i + \delta_t$ is also considered where γ_i and δ_t are random variables expressing the individual effect and time effect of the changes of β_{it} respectively.

3 Prediction of future returns and risk

Recall that a market model is an identity when no specification is made on the error term ε_{it}. This feature allows us to argue as follows. It is natural to think that r_{it} will depend on some fundamentals variables such as accounting variables (e.g., PER, BER etc.) and macroeconomic variables and r_{mt} will depend on macro-economic variables. Hence the joint distribution of (r_{it}, r_{mt}) will depend on the fundamental variables possibly through the means, variances and covariances. In other words, we may specify the distribution as follows; conditional on some fundamentals variables, say $z_t = (z_{1t}, \cdots, z_{qt})'$

$$\begin{pmatrix} r_{it} \\ r_{mt} \end{pmatrix} \sim N(\begin{pmatrix} \mu_{it} \\ \mu_{mt} \end{pmatrix}, \begin{pmatrix} \sigma_{iit} & \sigma_{imt} \\ \sigma_{mit} & \sigma_{mmt} \end{pmatrix})$$

where the unknown parameters μ's and σ's may depend on z_t. Then the market model obtained from this distribution as an identity depends on z_t in the parameter α_{it}, β_{it} and ϕ_{iit}. Consequently we can specify a time-varying coefficient market model by specify-ing α_{it}, β_{it} and ϕ_{iit} as functions of the fundamental variables:

$$\alpha_{it} = \alpha_i(z_t), \quad \beta_{it} = \beta_i(z_t) \quad \text{and} \quad \phi_{iit} = \phi_{ii}(z_t).$$

This approach has been taken in Rosenberg and McKibben (1973), and this argument has been made in Chapter 3 for some nonlinear time series models.

In this section, we describe the model proposed by Rosenberg and McKibben (1973), which is a systematic variation model for parame-ters. By making an efficient use of annual fundamental (account-ing) data of firms as well as the history of stock prices, they attempt to predict one-year ahead expected return and risk or the probability distribution of returns in the framework of a market model. The use of a market model is viewed as a measure of central tendency for each security. In other words, the returns of individ-

ual stocks are considered to be pulled back to or to be regressed
on the market returns eventually so that it will be expressed as

$$(3.1) \quad r_{it} = \alpha_{it} + \beta_{it} r_{mt} + \varepsilon_{it},$$
$$E(\varepsilon_{it}) = 0, \quad \mathrm{Var}(\varepsilon_{it}) = \phi_{iit} = \phi_{it}^2$$

where $r_{it} = \log S_{it}/S_{it-1}$. As is discussed in Section 1, this
equation is an identity for each t as it stands. Hence ε_{it} is
simply a non-market factor in (3.1). The central tendency of individual return r_{it} to the market return r_{mt} is implemented by

$$(3.2) \quad \mathrm{Cov}(\varepsilon_{it}, \varepsilon_{js}) = \begin{cases} \phi_{it}^2 & \text{if } i = j \text{ and } t = s \\ 0 & \text{otherwise.} \end{cases}$$

In this specification of non-market factors ε_{it}'s, they are serially and corss-serially uncorrelated though the variance ϕ_{it}^2 will be
specified to be fluctuated with accounting data of each firm. Also
they are cross-sectionally uncorrelated, i.e., $\mathrm{cov}(\varepsilon_{it}, \varepsilon_{kt}) = 0$
($i \neq k$). Hence ϕ_{it}^2 under (3.2) is considered the risk specific
to the i-th stock. In this sense the formulation so far is rather
in the framework of the traditional CAPM. However, Rosenberg and
McKibben (1973) importantly take a view that the variance σ_{it}^2 of
r_{it} as well as β_{it} change over time, which they say is based on
the evidence of Rosenberg (1973d) that the fluctuations in variance
are predictable and that the distribution of r_{it} is nearly normal
at each t under a suitable specification of variations of the
variance. This observation will be consistent with the leptokurtic
nonnormality observed in Chapter 2 but seems to have been neglected.

The variability of β_{it} and ϕ_{it}^2 are specified as

$$(3.3) \quad \begin{aligned} \beta_{it} &= \gamma_1 w_{1it} + \cdots + \gamma_J w_{Jit} + \xi_{it} = \gamma' w_{it} + \xi_{it} \\ \phi_{it}^2 &= \delta_1 w_{1it} + \cdots + \delta_J w_{Jit} + \nu_{it} = \delta' w_{it} + \nu_{it} \end{aligned}$$

where $\alpha_{it} = \alpha$, $\gamma = (\gamma_1, \cdots, \gamma_J)'$, $\delta = (\delta_1, \cdots, \delta_J)'$, $w_{it} = (w_{1it},$
$\cdots, w_{Jit})'$ and w_{Jit}'s are some fundamentals variables in account-
ing data and macroeconomic data, which they refer to as "descrip-

tors". This specification of parameter change may be interpreted as the systematic variation stated in Section 2. The coefficients γ_j's and δ_j's are common to all the stocks. The differences of β_{it}'s and ϕ_{it}'s come from those of the fundamentals w_{ijt}'s.

The error terms ξ_{it} and ν_{it} are assumed to satisfy

(3.4) i) $\{\varepsilon_{it}\}$, $\{\xi_{it}\}$ and $\{\nu_{it}\}$ are independent,

    ii) ξ_{it}'s are iid with $E(\xi_{it})=0$ and $Var(\xi_{it})=\omega$
       for $i=1,\cdots,N$ and $t=1,\cdots,T$,

    iii) (a) ν_{it}'s are iid with $E(\nu_{it})=0$ and $Var(\nu_{it})=\theta$
         for $i=1,\cdots,N$ and $t=1,\cdots,T$ or

      (b) $\nu_{it}\equiv\nu_i$ is constant for each i.

Under the assumptions (3.2), (3.3) and (3.4) the market model (3.1) is expressed as a factor model;

(3.5) $\quad r_{it} = \alpha+(\Sigma_{j=1}^{J}\gamma_j w_{jit}+\xi_{it}) r_{mt}+\varepsilon_{it}$
$\qquad\quad = \alpha+\gamma_1 z_{1it}+ \cdots +\gamma_J z_{Jit}+u_{it}$ with

(3.6) $\quad z_{jit} = w_{jit} r_{mt}$ and $\quad u_{it} = \xi_{it} r_{mt}+\varepsilon_{it}.$

Hence conditional on $\{r_{mt}\}$, $\{w_{jit}\}$ and $\{\nu_{it}\}$,

(3.7) $\quad Var(u_{it})=\omega r_{mt}^2+\phi_{it}^2$, $\quad Cov(u_{it}, u_{ks})=0$ $(i\neq k)$,
$\qquad Cov(u_{it}, u_{is})=\omega r_{mt} r_{ms}$ $(t\neq s)$.

The intercepts are assumed to be a common constant for all the stocks because the differences may be explained by those of the fundamentals. Hence without loss of generality we can assume $\alpha\equiv 0$, otherwise take $z_{1it}\equiv 1$ for all i and T. To express the model in terms of vectors and matrices, let

(3.8) $\quad y_i=(y_{i1},\cdots,y_{iT})'$ and $\quad Z_i=(z_{i1},\cdots,z_{iT})'$,

where $z_{it}=(z_{1it},\cdots,z_{Jit})' : J\times 1$ and $u_i=(u_{i1},\cdots,u_{iT})' : T\times 1$. Then the model (3.5) is expressed as a model of N regression equations;

(3.9) $y_i = Z_i \gamma_i + u_i$ $(i=1, \cdots, N)$.

Further letting

$$y = \begin{pmatrix} y_1 \\ \cdot \\ \cdot \\ \cdot \\ y_N \end{pmatrix}: NT \times 1, \quad Z = \begin{pmatrix} Z_1 \\ \cdot \\ \cdot \\ \cdot \\ Z_N \end{pmatrix}: NT \times J, \quad u = \begin{pmatrix} u_1 \\ \cdot \\ \cdot \\ \cdot \\ u_N \end{pmatrix}: NT \times 1,$$

it is written as a single regression equation;

(3.10) $y = Z\gamma + u$.

Here the covariance matrix of u conditional on $\{r_{mt}\}$, $\{w_{ijt}\}$ and $\{v_{it}\}$ is

(3.11) $\Delta \equiv \mathrm{Cov}(u) = \mathrm{dia}\{\Delta_1, \cdots, \Delta_N\}: TN \times TN$ with

(3.12) $\Delta_i = \omega r_m r_m' + \Phi_i$,

(3.13) $\Phi_i = \mathrm{diag}\{\phi_{ii1}, \cdots, \phi_{iiT}\}: T \times T$ and $r_m = (r_{m1}, \cdots, r_{mT})'$.

If we take ⅲ)(b) in (3.4), from (3.3)

(3.14) $\phi_{iit} = \delta' z_{it} + v_i$ with v_i's constant.

Therefore the unknown parameters are estimated by the GLS (generalized least squares) method minimizing $(y - Z\gamma)' \Delta^{-1} (y - Z\gamma)$, which gives the GLSE of the form for γ :

(3.15) $\hat{\gamma} = (\sum_{i=1}^{N} Z_i' \hat{\Delta}_i^{-1} Z_i)^{-1} (\sum_{i=1}^{N} Z_i' \hat{\Delta}_i^{-1} y_i) \equiv \hat{\gamma}(\hat{\Delta})$

Of course, assuming normal distribution $u \sim N(0, \Delta)$, the MLE (maximum likelihood estimator) is obtained. It is noted that the MLE of γ is also of the form (3.15) as in Kariya and Toyooka (1985).

However, Rosenberg and McKibben (1973) assume $\xi_{it} \equiv \xi_i$ and v_{it}

$\equiv \nu_i$ for $\{\xi_{it}\}$ and $\{\nu_{it}\}$ and view that ξ_i and ν_i are realized at $t=1$. This implies that ξ_i and ν_i may be treated as unknown parameter as in the case ⅲ) (b) in (3.4) once they are realized at $t=1$. But they treat ξ_i and ν_i as random variables before data is generated and when they consider sample properties of the estima-tors they proposed (see their paper).

In their empirical analysis, the following 32 descriptors of individual securities are considered.

Accounting-based descriptors
1.* Standard deviation of a per-share earnings growth measure
2. Accounting beta, or covariability of earnings with overall corporate earnings
3. Latest annual proportional change in per-share earnings
4.* Dividend payout ratio (5)
5.* Logarithm of mean total assets (5)
6.* Standard & Poor's quality rating
7. Estimated probability of default on fixed payments
8.* Liquidity (the quick ratio)
9.* Absolute magnitude of per-share dividend cuts (5)
10.* Mean leverage (senior securities/total assets) (5)
11. Smoothed operating leverage (fixed charges/operating income)
12. Standard deviation of per-share operating income growth (5)
13. Growth measure for per-share operating income (5)
14. Operating profit margin
15. Retained earnings per dollar of total assets (5)
16. Growth measure for total assets (5)
17. Growth measure for total net sales (5)
18.* Growth measure of per-share earnings available for common (5)
19. Nonsustainable growth estimate
20.* Gross plant per dollar of total assets

Market-based descriptors
21.* Historical beta, a regression of stock return on market return over preceding calendar years in the sample, assuming alpha equals zero
22.* Standard error of residual risk (deviations from regression (21)) (5)
23. Marketability, measured as ratio of annual dollar volume of trading to mean annual dollar volume for all securities
24. Negative semi-deviation of returns (5)
25.* Share turnover as a percentage of shares outstanding
26.* Logarithm of unadjusted share price
27. Dummy variable equal to one if stock is listed on the NYSE in latest period, equal to zero, otherwise

Market-valuation descriptors
28. Smoothed dividend yield
29. Earnings/price ratio
30. Book value of common equity par share-price
31. Estimates of misvaluation based on naive growth forecasts
32.   (G) and (G2)

Here data is annual and (5) denotes five years average values. The sample periods are (1) 1954~1966 for 558 firms and (2) 1954~1970 for 578 firms. Descriptors are dropped off from regression equation when they are insignificant at 90% level, so that 13 descriptors with * are used. Their observations are summarized as follows.

a) $\hat{\alpha}$'s are not significant.

b) $\hat{\beta}_{it}$'s are very significant relative to the benchmark value $\beta_{it} \equiv 1$.

c) There are many negative estimates $\hat{\omega}$, for which $\hat{\omega} = 0$ is set.

d) Most t values of $\hat{b}_j$'s are significant.

e) Regression results for σ^2 are rather stable.

f) The systematic risk $\hat{\theta}$ and the specific risk $\hat{\nu}_n$ are rather small, implying an increase in predictive power for risk.

They also tested the predictive power of the model based on the sample for 1958~1966. Using the prediction MSE (mean square errors) for 1967~1970, they found that the predictive beta

$$\hat{\beta}_{it} = \Sigma_{j=1}^{J} \hat{b}_j w_{itj} \quad (i = 1, \cdots, N: \quad t = 1967 \sim 1970)$$

with $\hat{\alpha}_{it} \equiv 0$ reduces the MSE relative to other β's such as historical beta.

4 A model in Japan

The basic idea discussed in Section 3 seems to be realized in what is known as related to the Nikko model in Japan. The main part of this section is based on Takahashi (1988). The model is a random coefficient model which can be reduced to a heteroscedastic regression model as in the model of Rosenberg and McKibben (1973) (see Section 3). In the model it is assumed that the return r_{it} of the i-th stock at time t follows the random coefficient model;

$$(4.1) \quad r_{it} = z_{i1t} f_{1t} + \cdots + z_{iqt} f_{qt} + \varepsilon_{it}.$$

($i=1, \cdots, p$: $t=1, \cdots, T$) (see (3.5) for comparison) where z_{ijt}'s are observable fundamentals and accounting variables which may be specific to each firm, and f_{jt}'s are the random coefficients. The list of "descripters" z_{ijt}'s in the model is given later.

Assumption

i) For $f_t = (f_{1t}, \cdots, f_{qt})'$: $q \times 1$, f_t's are iid with mean $E(f_t)$ $= \eta$ and covariance matrix $\text{Var}(f_t) = \varDelta$.

ii) For $\varepsilon_t = (\varepsilon_{1t}, \cdots, \varepsilon_{pt})'$: $p \times 1$, ε_t's are iid with $E(\varepsilon_t) = 0$ and $\text{Var}(\varepsilon_t) = \Psi = \text{diag}\{\psi_1, \cdots, \psi_p\}$.

iii) $\{f_t\}$ and $\{\varepsilon_t\}$ are uncorrelated processes; $\text{Cov}(f_t, \varepsilon_s) = 0$ ($t, s=1, \cdots, T$).

Estimation

The following estimation procedure is not in Takahashi (1988), but my procedure given the model. To rewrite the model (4.1), let

$$(4.2) \quad \underset{p\times1}{y_t} = \begin{pmatrix} r_{1t} \\ \cdot \\ \cdot \\ \cdot \\ r_{pt} \end{pmatrix} \quad \text{and} \quad \underset{p\times q}{Z_t} = \begin{pmatrix} z_{11t} & \cdots & z_{1qt} \\ \cdot & & \cdot \\ \cdot & & \cdot \\ \cdot & & \cdot \\ z_{p1t} & \cdots & z_{pqt} \end{pmatrix}.$$

Then

$$(4.3a) \quad y_t = Z_t f_t + \varepsilon_t = Z_t \eta + \xi_t \quad \text{with}$$

$$(4.3b) \quad \xi_t = Z_t(f_t - \eta) + \varepsilon_t.$$

Here since $E(\xi_t) = 0$ and

(4.4) $\mathrm{Var}(\xi_t) \equiv \Omega_t = Z_t \Delta Z_t{}' + \Psi,$

the model (4.3) is a regression model with heteroscedastic covariance matrix Ω_t for each t. Further, letting

$$(4.5) \quad y = \begin{pmatrix} y_1 \\ \cdot \\ \cdot \\ \cdot \\ y_T \end{pmatrix}, \quad Z = \begin{pmatrix} Z_1 \\ \cdot \\ \cdot \\ \cdot \\ Z_T \end{pmatrix} \quad \text{and} \quad \xi = \begin{pmatrix} \xi_1 \\ \cdot \\ \cdot \\ \cdot \\ \xi_T \end{pmatrix},$$

then the total observations are put in the form

(4.6) $y = Z\eta + \xi$

with ξ having the block diagonal covariance matrix;

(4.7) $\mathrm{Var}(\xi) = \mathrm{diag}\{\Omega_1, \cdots, \Omega_T\} \equiv \Omega.$

An efficient estimation procedure for such a model will be as follows. First, the regression coefficient vector η is estimated by the OLSE (ordinary least squares estimator)

(4.8) $\hat{\eta} = (Z'Z)^{-1} Z' y$

via (4.6). Second, by (4.4) f_t is estimated as the OLSE

(4.9) $\hat{f} = (Z_t' Z_t)^{-1} Z_t' y_t$

so that the covariance matrix Δ of f_t is estimated as

(4.10) $\hat{\Delta} = \frac{1}{T} \Sigma_{t=1}^{T} (\hat{f}_t - \hat{\eta})(\hat{f}_t - \hat{\eta})'.$

On the other hand, estimating ε_t by the OLS residual

(4.11) $\hat{\varepsilon}_t = y_t - Z_t \hat{f}_t,$

the diagonal elements ϕ_i of $\Psi = \mathrm{Var}(\varepsilon_t)$ is estimated as

(4.12) $\hat{\phi}_i = \frac{1}{T} \Sigma_{t=1}^{T} \hat{\varepsilon}_{it}^2.$

Thus we obtain an estimate of the covariance matrix Ω_t of ξ_t by

(4.13) $\hat{\Omega}_t = Z_t \hat{\Delta} Z_t' + \hat{\Psi}.$

Third, we re-estimate η in (4.6) by applying the GLS (generalized least squares) method as

(4.14) $\tilde{\eta} = (Z' \hat{\Omega}^{-1} Z)^{-1} Z' \hat{\Omega}^{-1} y,$

where $\hat{\Omega}$ is Ω in (4.7) with Ω_i's replaced by $\hat{\Omega}_i$'s, and f_t as

(4.15) $\tilde{f}_t = (Z_t' \hat{\Omega}_t^{-1} Z_t)^{-1} Z_t' \hat{\Omega}_t^{-1} y_t.$

Finally Δ, ψ_i and Ω_t are respectively estimated as

$$\tilde{\Delta} = \frac{1}{T} \Sigma_{t=1}^{T} (\tilde{f}_t - \tilde{\eta})(\tilde{f}_t - \tilde{\eta})'$$

$$\tilde{\psi}_i = \frac{1}{T} \Sigma_{t=1}^{T} \hat{\varepsilon}_{it}^2 \quad \text{and} \quad \tilde{\Omega}_t = Z_t \tilde{\Delta} Z_t' + \tilde{\Psi}.$$

As descripters z_{ijt}'s, it is reported that the following measures are taken into account.

(1) Systematic variability, (2) Specific variability, (3) Trading activity, (4) Success (performance of each stock relative to the average of an industry and a market), (5) Relative price momentum (which uses alphas), (6) Size of firm (market value and total asset value), (7) Growth, (8) Value to price, (9) Sales revenue to price, (10) Annual dividend yield, (11) Export revenue, (12) Financial leverage.

In addition to these descripters, 30 industry index factors are used. In other words, $q=42$ descripters are used in total.

Construction of portfolio

In forming a portfolio based on the model (4.3), the CAPM type concept is used to evaluate the forecasted return and risk of a portfolio. To describe it, let one-period ahead returns of a market portfolio and of a special portfolio be

(4.17) $\quad r_{Mt} = d_{Mt}' \ y_t \quad$ with $\quad d_{Mt} = (d_{Mt1}, \cdots, d_{Mtp})'$, and

(4.18) $\quad r_{At} = d_{At}' \ y_t \quad$ with $\quad d_{At} = (d_{At1}, \cdots, d_{Atp})'$

respectively. Since $(r_{Mt}, \ r_{At})$ is a bivariate random vector, by the argument in Section 1 we obtain the identity

(4.19) $\quad r_{At} = \alpha_{At} + \beta_{At} \ r_{Mt} + u_{At} \quad$ with
$$\mathrm{Cov}(u_{At}, \ r_{Mt}) = 0 \quad \text{and} \quad E(u_{At}) = 0, \quad \text{where}$$

(4.20) $\quad \alpha_{At} = E(r_{At}) - \beta_{At} E(r_{Mt})$
$$= d_{At}' \ Z_t \eta - \beta_{At} d_{Mt}' \ Z_t \eta,$$

(4.21) $\quad \beta_{At} = \dfrac{\mathrm{Cov}(r_{At}, \ r_{Mt})}{\mathrm{Var}(r_{Mt})} = \dfrac{d_{At}' \ \Omega_t d_{Mt}}{d_{Mt}' \ \Omega_t d_{Mt}}, \quad$ and

(4.22) $\quad u_{At} = r_{At} - \alpha_{At} - \beta_{At} \ r_{Mt}$
$$= (d_{At} - \beta_{At} d_{Mt})' \ Z_t (f_t - \eta) + (d_{At} - \beta_{At} d_{Mt})' \ \varepsilon_t$$
$$\equiv u_{ct} + u_{st}.$$

Hence the "unsystematic" risk in the market model (4.19) is divided into two parts;

(4.23) $\quad \mathrm{Var}(u_{At}) = \sigma_{ct}^2 + \sigma_{st}^2 \quad$ with

(4.24) $\quad \sigma_{ct}^2 = (d_{At} - \beta_{At} d_{Mt})' \ Z_t \Delta Z_t' (d_{At} - \beta_{At} d_{Mt}),$
$$\sigma_{st}^2 = (d_{At} - \beta_{At} d_{Mt})' \ \Psi(d_{At} - \beta_{At} d_{Mt}).$$

Thus the portfolio risk is evaluated as

(4.25) $\quad \sigma_{At}^2 = \beta_{At}^2 \sigma_{Mt}^2 + \sigma_{ct}^2 + \sigma_{st}^2.$

Since Ω_t in (4.21) and Δ and Ψ in (4.24) are estimated in the above method, the undetermined parameter in (4.25) is the portfolio value ratio vector d_{At}. Consequently, following the procedure described in Chapter 6, we can construct an optimal portfolio d_0 by maximizing the following objective function

(4. 26)

$$\Phi(d)=(d'\, Z_t\, \hat{\eta})^2-\{\lambda_M \beta_{Mt}^2(d)\, \sigma_{Mt}^2+\lambda_c \sigma_{ct}^2(d)+\lambda_s \sigma_{st}^2(d)\}$$

under certain conditions. Here the predicted one-period ahead return and risk are $\hat{\mu}_{At}=d_0'\, Z_t\, \hat{\eta}$ and $\hat{\sigma}_{At}^2 \equiv \hat{\sigma}_{At}^2(d_0)$ respectively, in which Z_t must be predicted at time $t-1$.

5 Convergent parameter model

Another time-varying coefficient model proposed by Rosenberg (1973c) is a regression model in which the coefficients tend to con verge to a cross-sectional mean value at each time t. He rationalizes such a model:

"When conformity is highly valued, or when the role of a deviate is, for any reason, difficult, individuals will tend to converge in behavior and in environment toward group norms, or toward subgroup norms if a deviant subgroup coalesces."

Incidentally this is in many respects what the Japanese society is. On the other hand, he also describes some factors to disturb such a converging trend. In particular, exogenous stochastic shocks are one of the factors. To fix the idea, let us consider the following regression model

(5. 1) $r_{i1} = b_{1it}\, f_{1it}+ b_{2it}\, f_{2it}+ u_{it}$ $(i=1,\cdots, N)$

with parameter variation

(5. 2) $b_{kit+1} = \bar{b}_{kt}+ \psi_k(b_{kit}-\bar{b}_{kt})+ \eta_{kit}$ $(k=1,2),$

where $\bar{b}_{kt}$ is the cross-sectional (group) mean of b_{kit}'s;

(5. 3) $\bar{b}_{kt} = \frac{1}{N}\Sigma_{i=1}^N b_{kit}$

and ψ_k $(0<\psi_k<1)$ denotes the rate of convergence to the group mean $\overline{b}_{kt}$ for each t ($t=1, \cdots, T$). It is noted that ψ_k is common to all stocks ($i=1, \cdots, N$) and that the mean $\overline{b}_{kt}$ changes over time so that (5.2) simply describes a stochastic tendancy of individual coefficient b_{kit} converging to the mean $\overline{b}_{kt}$ at each time. In other words, unless $\overline{b}_{kt}$ converges over time, b_{kit} will not converge. To validate the model (5.1), write it as

(5.4) $y_t = Z_t \beta_t + u_t$ where

(5.5)

$$
y_t = \begin{pmatrix} r_{1t} \\ \cdot \\ \cdot \\ \cdot \\ r_{Nt} \end{pmatrix}, \quad Z_t = \begin{pmatrix} z_{1t}{}' & 0 \\ & \\ 0 & z_{Nt}{}' \end{pmatrix}, \quad \beta_t = \begin{pmatrix} b_{1t} \\ \cdot \\ \cdot \\ b_{Nt} \end{pmatrix}
$$

$$
b_{it} = \begin{pmatrix} b_{1it} \\ b_{iit} \end{pmatrix}, \quad z_{it} = \begin{pmatrix} f_{1it} \\ f_{2it} \end{pmatrix} \quad \text{and} \quad u_t = \begin{pmatrix} u_{1t} \\ \cdot \\ \cdot \\ \cdot \\ u_{Nt} \end{pmatrix}.
$$

Also write (5.2) as

(5.6) $\beta_{t+1} = \overline{\beta}_t + (I_N \otimes D_\phi)[\beta_t - \overline{\beta}_t] + \eta_t$,   where

(5.7) $\overline{\beta}_t = 1 \otimes \overline{b}_t$ with $\overline{b}_t = (\overline{b}_{1t}, \overline{b}_{2t})'$,

$D_\phi = \text{diag}\{\psi_1, \psi_2\}$ and $\eta_t = (\eta_{1t}, \cdots, \eta_{Nt})$ with $\eta_{it} = (\eta_{1it}, \eta_{2it})'$. Here $1 = (1, \cdots, 1)' : N \times 1$. Further (5.6) is expressed as

(5.8) $\beta_{t+1} = B\beta_t + \eta_t$ with

(5.9) $B = I_N \otimes D_\phi + (M \otimes I_N)[I_{2N} - I_N \otimes D_\phi]$, and

where $M = 1(1'1)^{-1}1'$. Hence the model (5.1) with (5.2) is equivalent to

(5.10) $y_t = Z_t \beta_t + u_t$,   and $\beta_{t+1} = B\beta_t + \eta_t$, which is a state space model discussed in Chapter 4.

Rosenberg made the following assumptions for error terms;

(5.11) $\quad \begin{pmatrix} u_t \\ \eta_t \end{pmatrix} {}^N_{2N} \quad \sim \quad iid \quad N(0, \begin{pmatrix} \Omega & 0 \\ 0 & \Phi \end{pmatrix}), \quad$ where

(5.12) $\quad \Omega = (\omega_{ii}) = \gamma \begin{pmatrix} 1 & \lambda \\ \lambda & 1 \end{pmatrix}, \quad$ and

(5.13) $\quad \Phi = (\phi_{ij}) = \delta \begin{pmatrix} I_2\Delta & \cdots & \Delta \\ & \cdot & \\ & \cdot & \\ \Delta & \cdots\cdots & I_2 \end{pmatrix}$

with $\Delta = \mathrm{diag}\{\tau_1, \tau_2\}$. Here the off-diagonal ellements in Ω are all equal to λ and the 2×2 off-diagonal block matirces in Φ are all equal to Δ. Therefore the model (5.10) with (5.11) fits the framework of the Kalman filtering procedure described in Chapter 4. Consequently once the unknown parameter vector

$$\theta = (\gamma, \lambda, \delta, (\tau_1, \tau_2))$$

are estimated, we can obtain the predictive beta's at time $t+1$, $t+2, \cdots$ on which a portfolio can be made or a group of stocks or assets can be selected.

Exercises

1. Show $\mathrm{Cov}(x, \varepsilon) = 0$ in (1.3).

2. Show (1.7) and (1.8).

3. Give an example in which a random dispersed parameter model will be effective.

4. Prove (4.20), (4.21) and (4.24).

5. Derive (5.6) and (5.8).

CHAPTER 9

SELECTION OF PORTFOLIO POPULATION

1 Introduction

As has been discussed in Chapter 6, to construct a portfolio it is important to select a set of financial assets in advance from a universe of available assets. Criteria for selection of a portfolio population must correspond to investment preferences or stances. But an investment stance needs to be feasible relative to available financial instruments and to be desirably consistent with their time series variational features though we may not have enough knowledge about them. On the other hand, whatever instruments may be, an investment stance should be expressed in terms of future returns and risks of a portfolio to be constructed (see Chapter 6), and no doubt, a portfolio with higher predicted returns and lower predicted risks is more preferable. In other words, a procedure of selecting a portfolio population is a procedure of predicting future returns and risks of many assets as a whole and of picking a class of assets whose portfolio is predicted to perform well relative to an investment stance. Needless to say, available assets are too enormous to treat simultaneously and besides prediction procedures depend on financial instruments. Hence a first step is to treat each instrument separately.

A second step is to choose an approach to prediction and a scope of data. As has been stated in Chapters 1 and 6, there are three groups of analysts who forecast future returns and risks of individual assets;

1) "trads" who engage in fundamental analysis,

2) "techs" who engage in technical analysis, and

3) "quants" who engage in model analysis.

"Trads" base their predictions on financial statement analysis, company visits, analytic insights etc., and "techs" use such concepts

as price trend and market sentiment with expertized chart techniques. While, "quants" use statistical models with computers.
The role of these analysts in asset allocation is commonly to
select a set of assets and construct a portfolio with higher
performance. They commonly use data, but they use it differently.
Questions to pose here are

(1) what kind of data and what form of data really contains predictive information about future returns and risks, and

(2) how long it takes to select an optimal portfolio population
from a large number of assets.

The question (1) is hard to answer. Analysts in each group think
that their way of use of data is "relevant" to predicting returns
and risks. But for the question (2), a quants approach will be a
promising answer because, for example, there are about 1,500
stocks listed in the Tokyo Stock Market. Whatever approach may be
taken, data set consists of

a) price data $\{ S_{it} \}$ ($i=1, \cdots, N$: $t=1, \cdots, T$)   and

b) ancillary (fundamental) data $\{ z_{jt} \}$ ($j=1, \cdots, M$: $t=1, \cdots, T$).

Here the length of sample period needs to be specified in advance
relative to an investment horizon or the purpose of analysis. We
also need to specify forms of variates in data. It should be noted
that even if we are interested in prediction of returns and risks,
it does not mean that we should analyze return series. In fact, as
has been pointed out in Chapters 3 and 6, analysis of returns implies analysis of high frequency variations of prices and it is
likely to be influenced by erratically small errors. A suggested
form of price variable in asset allocation may be $\tilde{x}_t = S_t / S_T$
though it will be heteroscedastic (see Chapter 6).

In this chapter, we consider two procedures of using an ancillary set of variables which will have predictive power for selection of a portfolio population. Specifically we describe the first
two of the following three quants approaches to selection of a

portfolio population;

1.   IC (information coefficient) model approach (Jones (1987)),

2.   Classification analysis approach,

3.   Cluster analysis approach.

Cluster analysis is discussed in association with index (tracking) portfolio in Chapter 11. Of course, before such an approach is taken, the total population may be filtered by a screening system. In Chapter 10, a random grouping MTV approach will be presented for selection of stocks where no ancillary variables are used.

2 Information coefficient (IC) for selection of a portfolio population

When an analyst selects some stocks from a universe to make a portfolio, he will analyze various data through his expertized knowledge and his model in the head and select some stocks whose prices are predicted to rise. The personal analytical model in his head which cannot be shared with other people is in fact his value and may be flexible for the evolving variational structure of stock prices. On the other hand, the model will be rather descriptive and judgemental. Typically an analyst rests his evaluation of stocks on the following criteria;

(2.1)   ① value, ② yield, ③ momentum, ④ growth,
        ⑤risk, ⑥liquidity

etc. Jones (1987) viewed the stock-selecting procedure of an analyst as a "multifactor" model;

(2.2) $a_1①+ a_2②+ \cdots + a_6⑥$ where a_i's are weights,

and developed a computer-generated stock selection system as opposed to the analyst-derived stock selection procedure. In this section, we briefly introduce his procedure together with the so-called IC (information coefficient), which is rank-order correlation coefficient in statistics.

As variables representing the concepts in (2.1), Jones (1987) chose for ① value (1) DDM (dividend discount model) and (2) K ratio, for ② yield (3) cash flow yield and (4) erning yield, for ③ momentum (5) earning momentum and (6) pricemomentum, for ④ growth (7) historical growth and (8) sustainable growth, for ⑤ risk (9) price volatility and (10) EPS (earning per share) uncertainty, and for ⑥ liquidity (11) capitalization and (12) analysts' coverage. For the details of these variables, the readers are referred to his original paper. To obtain a computer-generated stock selection model in (2.2), he evaluated the predictability of each variable in terms of rank correlation coefficients and combined the relative ranks.

IC — rank-order correlation coefficient

Let $z_i(j)$ be the j-th fundamental variable of the i-th stock where $i=1,\cdots,N$ and $j=1,\cdots,12$ as above. The N stocks are ranked in the ascending order of the size of $z_i(j)$ for given j. Here taking the inverse of $z_i(j)$ if necessary, without loss of generality we may regard that the larger the $z(j)$-value is, the more the stock is preferred. Then for given the j-th variable, the N stocks are ranked as

(2.3)
stocks	1	,	2	,	3	,	·············	,	N
ranks	$a_1(j)$,		$a_2(j)$,		$a_3(j)$,		·············	,	$a_N(j)$

where $a_i(j)$ is the rank of the i-th stock with respect to the size of the j-th variable. If the j-th variable has a predictive power for future returns of stocks, then the sizes of the returns after a certain period will match the ranks in (2.3) in their orders. In other words, letting x_{ki} be the return of the i-th stock after k periods, the returns x_{k1}, $x_{k2},\cdots,x_{kN}$ will have a higher correlation $r_k(j)$ with the ranks in (2.3) if the j-th variable is effective in prediction, where

(2.4) $$r_k(j) = \frac{\sum_{i=1}^{N}(a_i(j)-\overline{a}(j))(x_{ki}-\overline{x}_k)}{[\sum_{i=1}^{N}(a_i(j)-\overline{a}(j))^2 \sum_{i=1}^{N}(x_{ki}-\overline{x}_k)^2]^{1/2}}$$

with $\bar{a}(j)=\sum_{i=1}^{N} a_i(j)/N$ and $\bar{x}_k=\sum_{i=1}^{N} x_{ki}/N$. This rank-order correlation is often referred to as IC (information coefficient) in finance. Jones considered the $r_k(j)$ for $k=1,3,6,12,24$ (months) for each j and the average $\bar{r}(j)=\sum_k r_k(j)/5$. Based on $r_k(j)$, the predictability of the j-th variable for returns is checked for each future period k and at each time t.

To combine the predictive powers of all the 12 variables, Jones defined the relative rank of the i-th stock with respect to the j-th variable by $a_i(j)/N$ and the relative rank of the i-th stock by the 12 variables by

$$(2.5) \quad \bar{a}_i = \frac{1}{12}\sum_{j=1}^{12} a_i(j)/N \quad (i=1,\cdots,N).$$

Then the N stocks are ranked with respect to the sizes of $\bar{a}_i$'s;

$$(2.6) \quad \text{stocks} \quad 1 \quad , \quad 2 \quad , \cdots\cdots\cdots , \quad N$$
$$\text{ranks} \quad b_1(\bar{a}), \quad b_2(\bar{a}), \quad \cdots\cdots\cdots , \quad b_N(\bar{a})$$

where $b_i(\bar{a})$ is the rank of the i-th stock with respect to the sizes of $\bar{a}_i$'s. Again the predictability of measure $\bar{a}_i$ in (2.5) is measured by the rank-order correlation coefficient $r_k(\bar{a})$, which is $r_k(j)$ in (2.4) with $a_i(j)$ replaced by $b_i(\bar{a})$.

3 Classification procedure

A well specified IC model will select a set of stocks whose portfolio performs well. But this does not mean that an IC model can predict future returns of individual stocks with desired accuracy. In fact, ancillary variables z's may not carry sufficient information to predict individual future returns, even though they carry enough information to pick stocks which will perform well as a whole or as a portfolio. In other words, the variables may contain enough information to discriminate "good" stocks from "bad" stocks, suggesting a use of classification method, another multivariate

statistical method. In the next section, a general robust classification procedure is developed. In this section, we describe a
direct classification procedure as a method for selecting a
portfolio population. Classification analysis is often referred to
as discriminant analysis.

Let S_{it} be the price of the i-th asset at time t where $i=1$,
$\cdots$, N, and let $\{x_{it}\}$ be a transformed series of $\{S_{it}\}$ for discrimination such as

$$x_{it} = \log S_{it} - \log S_{it-1} \quad \text{or} \quad \widetilde{x}_{it} = S_{it} / S_{iT},$$

where $\{t=1, \cdots, T\}$ is a sample period. Also let $(z_{i1t}, \cdots, z_{iMt})$
be ancillary variables selected in advance in some manner where
z_{ijt}'s $(j=1, \cdots, M)$ are variables associated with the i-th asset.
A statistical procedure for selecting ancillary variables will be
also discussed later. Since the variables z's as a whole are expected to be able to discriminate "goodies" (G) from "badies" (B),
a method to construct a classification function rule is to combine
them linearly as

$$(3.1) \quad D_t(i) = a_{1t} z_{i1t} + \cdots + a_{Mt} z_{iMt} = a_t' z_{it}$$

and find weights a_{jt}'s in order for $D_t(i)$ to efficiently discriminate (G) from (B), where $a_t = (a_{1t}, \cdots, a_{Mt})'$ and $z_{it} = (z_{i1t},$
$\cdots, z_{iMt})'$. Of course, an optimal classification rule may not be
linear as we will discuss soon. In our approach, we use cross-
sectional data to obtain a classification function so that the
parameters defining a classification function depend on t, which
will provide us a facility to forecast the future classification
function.

A procedure for obtaining a classification rule is as follows,
where t is fixed.

① Order the x_{it}'s in their sizes from the largest to the smallest. Without loss of generality, let them be

(3.2) $x_{1t} \geqq x_{2t} \geqq \cdots \geqq x_{Nt}.$

Then pick the top K stocks as (G) and the bottom K stocks as (B). Here K should be more than $2M$.

② Regard z_{it} as a random variable with mean and covariance matrix: $E(z_{it}) = \mu_{it}$ and $\mathrm{Var}(z_{it}) = \Sigma_{it}$. Let $G = \{1, \cdots, K\}$ and $B = \{N-K+1, \cdots, N\}$ be the index sets of (G) and (B) respectively. We may expect μ_{it}'s and Σ_{it}'s to be similar within each group of (G) and (B). Hence we assume

(3.3) $\begin{aligned} \mu_{it} &= \mu_t^G \quad \text{and} \quad \Sigma_{it} = \Sigma_t^G \quad \text{for all} \quad i \in G, \\ \mu_{it} &= \mu_t^B \quad \text{and} \quad \Sigma_{it} = \Sigma_t^B \quad \text{for all} \quad i \in B. \end{aligned}$

Otherwise consider the averages of μ's and Σ's over each group and regard them as those in (3.3). Then from (3.1) the means and variances of $D_t(i)$ are

(3.4G) $E[D_t(i)] = a_t' \mu_t^G, \ \mathrm{Var}(D_t(i)) = a_t' \Sigma_t^G a_t \quad \text{for} \quad i \in G,$

(3.4B) $E[D_t(i)] = a_t' \mu_t^B, \ \mathrm{Var}(D_t(i)) = a_t' \Sigma_t^B a_t \quad \text{for} \quad i \in B,$

which are representative characteristics of (G) and (B) under the linear classification rule. We may choose a_t in order to separate the characteristics of (G) in (3.4G) from those of (B) in (3.4B) as far as possible.

③ The parameters in (3.3) are estimated by

(3.5) $\begin{aligned} \hat{\mu}_t^H &= \frac{1}{K} \Sigma_{i \in H} z_{it} \qquad\qquad\qquad H = G \text{ or } B, \\ \hat{\Sigma}_t^H &= \frac{1}{K} \Sigma_{i \in H} (z_{it} - \hat{\mu}_t^H)(z_{it} - \hat{\mu}_t^H)' \quad H = G \text{ or } B. \end{aligned}$

Based on these estimates, test the hypothesis

(3.6) $H_0 : \mu_t^G = \mu_t^B, \ \Sigma_t^G = \Sigma_t^B$

with test statistic $R_{0t} = -2 L \log \lambda_t$, where

(3.7) $\lambda_t = \dfrac{(2k)^{Mk} |K \hat{\Sigma}_t^G|^{k/2} |K \hat{\Sigma}_t^B|^{k/2}}{k^{Pt} |A + K \hat{\Sigma}_t^G + K \hat{\Sigma}_t^B|}$, $k = K-1$ with

$$A= K(\hat{\mu}_t^G- \hat{\mu}_t)(\hat{\mu}_t^G- \hat{\mu}_t)' + K(\hat{\mu}_t^B- \hat{\mu}_t)(\hat{\mu}_t^B- \hat{\mu}_t)',$$

$$\hat{\mu}_t=\frac{1}{2}(\hat{\mu}_t^G+\hat{\mu}_t^B) \quad \text{and} \quad L=1-\frac{1}{k}\frac{M-k+1}{m+3}.$$

The λ_t in (3.7) is a modified LRT (likelihood ratio test) statistic under the assumption that z_{it}'s ($i \in G$) are iid $N(\mu_t^G, \Sigma_t^G)$,
z_{jt}'s ($j \in B$) are iid $N(\mu_t^B, \Sigma_t^B)$, and $\{z_{it} : i \in G\}$ and $\{z_j:$
$j \in B\}$ are independent. The distribution of R_{0t} is approximated
by χ^2-distribution with d.f. (degrees of freedom) $f=(k-1)M(M$
$+3)/2$ when $k= K-1$ is large, and the hypothesis H_0 is rejected
if $R_{0t}> c_\alpha$, where c_α satisfies $P(R_{0t}> c_\alpha)= \alpha$. Since z_{it}'s
may not satisfy the required normal assumption, the test is approxi-
mate. See Siotani, Hayakawa and Fujikoshi (SHF below) (1985, pp.
355-356) for details. If the hypothesis is not rejected, there
will be no chance to use z_{it}'s for discrimination.
④   When H_0 in (3.6) is rejected, test the hypothesis

$$(3.8) \quad H_1 : \Sigma_t^G= \Sigma_t^B \quad (\equiv \Sigma_t)$$

by the test statistic

$$(3.9) \quad R_{1t} = \frac{1}{2}\left[k\mathrm{tr}(\hat{\Sigma}_t^G\hat{\Sigma}_t^{-1}- I)^2+ k\mathrm{tr}(\hat{\Sigma}_t^B\hat{\Sigma}_t^{-1}- I)^2\right] \quad \text{with}$$

$$(3.10) \quad \hat{\Sigma}_t = \frac{1}{2}(\hat{\Sigma}_t^G+ \hat{\Sigma}_t^B).$$

This test has a local optimality. The distribution of R_{1t} is approximated by χ^2 distribution with d.f. $f=(k-1)M(M+1)/2$ and
H_1 is rejected when $R_{1t}> c_\alpha$. See SHF (1985) for details and
Kariya and Sinha (1989) for a robustness property of the test.

   If the hypothesis H_1 in (3.8) is not rejected, use the following linear discriminant function

$$(3.11) \quad D_t(i) = \lceil z_i-\frac{1}{2}(\hat{\mu}_t^G+ \hat{\mu}_t^B)\rceil' \hat{\Sigma}_t^{-1}(\hat{\mu}_t^G- \hat{\mu}_t^B),$$

where $\hat{\Sigma}_t$ is given by (3.10) and z_i is an object to be assigned to
G or B. The classification rule is;

$$(3.12) \quad \begin{array}{l} \text{assign} \quad i \text{ to } G \quad \text{if} \quad D_t(i) > 0, \\ \text{assign} \quad i \text{ to } B \quad \text{if} \quad D_t(i) \leq 0. \end{array}$$

The classification function (3.11) may be regarded as an estimated version either of Fisher's linear discriminating function which maximizes $|a_t' \mu_t^G - a_t' \mu_t^B| / a_t' \Sigma_t a_t$, as stated in ① or of a Bayes rule under the normal assumption stated in ③.

⑤ If the hypothesis $H_1: \Sigma_t^B = \Sigma_t^G$ in (3.8) is rejected, as will be suggested in the next section we may test the weaker hypothesis

$$\widetilde{H}_1 : \quad |\Sigma_t^B| = |\Sigma_t^G|.$$

A test statistic for this hypothesis is given in (4.26) of the next section. When $\widetilde{H}_1$ is not rejected, the following quadratic classification rule can be used:

$$(3.13)$$
$$D_t^*(i) = \frac{1}{2} z_i' [(\hat{\Sigma}_t^B)^{-1} - (\hat{\Sigma}_t^G)^{-1}] z_i + z_i' [(\hat{\Sigma}_t^G)^{-1} \hat{\mu}_t^G - (\hat{\Sigma}_t^B)^{-1} \hat{\mu}_t^B]$$
$$+ \frac{1}{2} [\hat{\mu}_t^{B'} (\hat{\Sigma}_t^B)^{-1} \hat{\mu}_t^B - \hat{\mu}_t^{G'} (\hat{\Sigma}_t^G)^{-1} \hat{\mu}_t^G],$$

no matter what the underlying distributions may be as far as they have certain elliptically contoured pdf's (see Section 4). The classification rule is (3.12) with $D_t(i)$ replaced by $D_t^*(i)$. On the other hand, if it is reasonable to assume multivariate normality for z_{it}'s, then without testing the hypothesis $\widetilde{H}_1$ we can use

$$(3.14) \quad D_{1t}^*(i) = D_t^*(i) + \log\{|\hat{\Sigma}_t^B| / |\hat{\Sigma}_t^G|\} / 2.$$

Before one of these rules is used, one may test the equality of means;

$$(3.15) \quad H_2 : \mu_t^G = \mu_t^B.$$

But since $\Sigma_t^G \neq \Sigma_t^B$, this testing problem is known as the Behrens-Fisher problem and is difficult to treat. If we sacrifice some information, we can test it by rejecting H_2 when

$$(3.16) \quad W_t = \frac{K-M}{M} \bar{y}_t' \ S_t^{-1} \bar{y}_t > c_\alpha,$$

where with $y_{it} = z_{it} - z_{N-it}$ ($i = 1, \cdots, K$),

$$\bar{y}_t = \frac{1}{K} \Sigma_{i=1}^K y_{it} \quad \text{and} \quad S_t = \frac{1}{K} \Sigma_{i=1}^K (y_{it} - \bar{y}_t)(y_{it} - \bar{y}_t)'.$$

The distribution of W_t is approximated by F-distribution with d.f. (degrees of freeedom) ($M, K-M$), which is exact under normality. If H_2 is rejected, we use (3.13). If H_2 is accepted, we may modify (3.14) under normality as

$$(3.17) \quad D_t^{**}(i) = \frac{1}{2}(z_i - \hat{\mu}_t)' [(\hat{\Sigma}_t^B)^{-1} - (\hat{\Sigma}_t^G)^{-1}] (z_i - \hat{\mu}_t)$$
$$+ \frac{1}{2} \log\{|\hat{\Sigma}_t^B| / |\hat{\Sigma}_t^G|\} \quad \text{with} \quad \hat{\mu}_t = \frac{1}{2}(\hat{\mu}_t^G + \hat{\mu}_t^B).$$

Evaluation of discriminant power

The power of the classification function $D_t(i)$ in (3.11) (or $D_t^*(i)$ in (3.13) or $D_{1t}^*(i)$ in (3.14) or $D_t^{**}(i)$ in (3.17)) can be tested as follows.

i) Insert z_{it}'s with $i \in G$ into $D_t(i)$ in (3.11) and classify them according to the rule (3.12) to get the number K_G of correctly assigned i's in G. Similarly get the number K_B of correctly assinged j's in B. Then the in-sample ratios of correct classification are $k_g = K_G/K$ and $k_b = K_B/K$ for each group, and $k_{gb} = (K_G + K_B)/2K$ is the total ratio of correct classification. The larger these ratios are, the better the function. Next, z_{it}'s in the top $2K$ i's and the bottom $2K$ j's of the ranking stated in 1 are inserted into the classification function and it is checked whether or not they are well classified. These tests are in-sample tests.

ii) As an out-of-sample test, z_{it+1}'s at $t+1$ are substituted into the discriminant function $D_t(i)$ at t, and in the same manner as in i), it is checked whether or not $D_t(i)$ can have a classifying power for the next period. Let the predictive discriminant ratios be denoted by

$$k_{g\,t+1\,|\,t}, \quad k_{b\,t+1\,|\,t} \quad \text{and} \quad k_{g\,b\,t+1\,|\,t}.$$

These ratios will be important in practice and their time series trend will enable us to judge the validity of an actual use of $D_t(i)$ at $t=T$.

Predictive MTV classification rule

To apply a classification rule to a practical investment decision at time T for $T+n$, we need

1) predicted values of z_{it}'s at $t=T+n$, and

2) predicted values of unknown parameters (μ_t^H, Σ_t^H) at $t=T+n$

to get a classification rule $D_{T+n}(i)$ where $H=G, K$.

In general, it is not easy to make predictions on $z_{i\,T+n}$'s. To avoid this prediction, one may follow the following procedure;

(1) order $x_{i\,T}$'s as in ① above, and

(2) use $z_{i\,T-1}$'s and follow the procedures from ② to ⑥ with

$z_{i\,T-1}$'s to get a classification rule.

For simplicity, we assume that the classification rule thus obtained is linear as in (3.11) and denoted by

$$(3.18) \quad D_{T\,|\,T-1}(i) = \hat{\eta}_{T\,|\,T-1}{}' z_i - \hat{\delta}_{T\,|\,T-1} \quad \text{with}$$

$$\hat{\eta}_{T\,|\,T-1} = \hat{\Sigma}_{T-1}^{-1}(\hat{\mu}_{T-1}^G - \hat{\mu}_{T-1}^B) \quad \text{and}$$

$$\hat{\delta}_{T\,|\,T-1} = \frac{1}{2}(\hat{\mu}_{T-1}^G + \hat{\mu}_{T-1}^B)' \hat{\eta}_{T\,|\,T-1}.$$

This is a predictive classification rule to discriminate (G) from (B) at T based on the data $z_{i\,T-1}$'s at $T-1$. If this rule is effective in the performance check described above, one may use $D_{T+1\,|\,T}(i)$ for given $z_{i\,T}$'s at T to predict on goodies and badies. But in that case we need $\hat{\eta}_{T+1\,|\,T}$ and $\hat{\delta}_{T+1\,|\,T}$, which are not available because $x_{i\,T+1}$'s have not been observed. A possibility to predict $\hat{\eta}_{T+1\,|\,T}$ and $\hat{\delta}_{T+1\,|\,T}$ is to use the MTV model for the data

$$\{(\hat{\pmb{\eta}}_{t|t-1}, \hat{\pmb{\delta}}_{t|t-1}) : t=2, \cdots, T\},$$

in which case $\hat{D}_{T+n|T+n-1}(i)$ is obtained ($n=1,2,\cdots$).

On the other hand, if z_{iT+n}'s are directly predictable, maybe again through an MTV model, a predictive $D_{T+n}(i)$ is obtained by predicting unknown parameters through an MTV model based on the data $\{(\hat{\pmb{\eta}}_{t|t}, \hat{\pmb{\delta}}_{t|t})\}$. There will be some other ways to combine cross-sectional and time series movements.

Selection of discriminating variables

A popular discriminant function is the one given in (3.11) where $\pmb{\Sigma}_t^G = \pmb{\Sigma}_t^B$ is assumed, which is an estimated version of the linear discriminant function

$$\pmb{a}_t' \pmb{z}_i - \frac{1}{2}(\pmb{\mu}_t^G - \pmb{\mu}_t^B)' \pmb{\Sigma}_t^{-1}(\pmb{\mu}_t^G + \pmb{\mu}_t^B)$$

with $\pmb{a}_t = \pmb{\Sigma}_t^{-1}(\pmb{\mu}_t^G - \pmb{\mu}_t^B)$. The significance of coefficients a_{it}'s for discriminating variables z_{it}'s are tested as follows. Let $\pmb{\delta}_t = \pmb{\mu}_t^G - \pmb{\mu}_t^B$,

$$\pmb{a}_t = \begin{pmatrix} \pmb{a}_{1t} \\ \pmb{a}_{2t} \end{pmatrix} \begin{matrix} m \\ M-m \end{matrix} \quad, \quad \pmb{\delta} = \begin{pmatrix} \pmb{\delta}_{1t} \\ \pmb{\delta}_{2t} \end{pmatrix} \begin{matrix} m \\ M-m \end{matrix}, \quad \text{and} \quad \pmb{\Sigma}_t = \begin{pmatrix} \pmb{\Sigma}_{11t} & \pmb{\Sigma}_{12t} \\ \pmb{\Sigma}_{21t} & \pmb{\Sigma}_{22t} \end{pmatrix} \begin{matrix} m \\ M-m \end{matrix}.$$

Then it follows that

$$\pmb{a}_{1t} = \pmb{\Sigma}_{11t}^{-1} \pmb{\delta}_{1t} - \pmb{B}_t' \pmb{\Sigma}_{22\cdot 1t}^{-1}(\pmb{\delta}_{2t} - \pmb{B}_t \pmb{\delta}_{1t}), \text{ and}$$
$$\pmb{a}_{2t} = \pmb{\Sigma}_{22\cdot 1t}^{-1}(\pmb{\delta}_{2t} - \pmb{B}_t \pmb{\delta}_{1t}),$$

where $\pmb{B}_t = \pmb{\Sigma}_{21t} \pmb{\Sigma}_{11t}^{-1}$ and $\pmb{\Sigma}_{22\cdot 1t} = \pmb{\Sigma}_{22t} - \pmb{\Sigma}_{21t} \pmb{\Sigma}_{11t}^{-1} \pmb{\Sigma}_{12t}$. Here we are interested in testing

$$H: \pmb{a}_{2t} = 0 \quad \text{or equivalently} \quad \pmb{\delta}_{2t} = \pmb{B}_t \pmb{\delta}_{1t}.$$

The test which rejects H for large values of

$$W_t = \frac{c[\pmb{d}_t'(2K\hat{\pmb{\Sigma}}_t)^{-1} \pmb{d}_t - \pmb{d}_{1t}'(2K\hat{\pmb{\Sigma}}_{11t})^{-1} \pmb{d}_{1t}]}{1 + (K/2) \pmb{d}_{1t}'(2K\hat{\pmb{\Sigma}}_{11t})^{-1} \pmb{d}_{1t}}$$

is an optimal test, where $c = K(2K-M-1)/2(M-m)$, $d_t = \hat{\mu}_t^G - \hat{\mu}_t^B$, and d_{1t} and $\hat{\Sigma}_{11t}$ correspond to δ_{1t} and Σ_{11t} respectively. The null distribution of W_t is approximated by F-distribution with d. f.'s $M-m$ and $2K-M-1$.

4 Robust classification analysis

In the development of classification method, it is usually assumed that variates are jointly distributed as a multivariate normal distribution. But in financial data the assumption may not be satisfied. In this section we develop a robust classification analysis procedure which does not heavily depend on the normal assumption.

Two population classification problem. To classify or assign a random observation x_0 into either of the known two parent populations;

$$(4.1) \quad \begin{matrix} \text{Population} & \Pi_1 & \text{with pdf} & f_1(x), \\ \text{Population} & \Pi_2 & \text{with pdf} & f_2(x), \end{matrix}$$

where pdf stands for probability density function and $f_i(x)$'s are known pdf's.

Since x_0 is a p-dimensional observation from either pdf f_1 or pdf f_2, it is a point in Euclidean space R^p. Hence as a classification rule (procedure), it is natural to consider the procedure of dividing R^p into two disjoint subsets A_1 and A_2 as

$$(4.2) \quad R^p = A_1 \cup A_2, \quad A_1 \cap A_2 = \phi$$

and classifying x_0 as follows;

$$(4.3) \quad \begin{matrix} \text{if} & x_0 \in A_1, & x_0 \text{ is judged to be an observation from } \Pi_1, \\ \text{if} & x_0 \in A_2, & x_0 \text{ is judged to be an observation from } \Pi_2. \end{matrix}$$

Then the classification problem is to decompose R^p into two disjoint subsets A_1 and A_2 in order that errors due to the rule (4.
3) may be made as small as possible. As x_0 is an random observation from either f_1 or f_2, the classification errors can be
evaluated as

(4.4)

the error probability $b(2|1)$ that we assign x_0 to Π_2
when x_0 is from Π_1,

the error probability $b(1|2)$ that we assign x_0 to Π_1
when x_0 is from Π_2,

where the error probabilities are given by

(4.5) $b(j|i) = \int_{A_j} f_i(x)\,dx$ $(i,\,j=1,2)$.

However, as the two error probabilities can not be made small simul
taneously, we choose A_1 and A_2 so that a linear combination of
the two probabilities

(4.6) $a_1 b(2|1) + a_2 b(1|2)$ with $a_1 + a_2 = 1$ and $a_1,\,a_2 \geqq 0$

is minimized. The weights a_i's are chosen in advance and are
often interpreted from a Bayesian viewpoint. The following lemma
is well-known.

Lemma   A division (A_1, A_2) of R^p in (4.2) minimizes (4.6) if
(and only if almost everywhere)

(4.7) $A_1 = \{x:\ f_1(x)/f_2(x) \geqq a_2/a_1\}$ and $A_2 = R^p - A_1$.

In most applications, p-dimensional normal distribution $N(\mu_i, \Sigma_i)$ is assumed for Π_i where the pdf of $N(\mu, \Sigma)$ with mean μ and
covariance matrix Σ is given by

(4.8)

$g(x:\mu,\Sigma) = (2\pi)^{-p/2} |\Sigma|^{-1/2} \exp(-\frac{1}{2}(x-\mu)' \Sigma^{-1}(x-\mu))$.

Also $a_1 = a_2 = 1/2$ is taken in many applications. In this section, to derive a robust classification rule, we replace the family of normal distributions by a class of elliptically symmetric distributions whose pdf's are of the form

$$(4.9) \quad h(x|\mu, \Sigma) = c|\Sigma|^{-1/2} q((x-\mu)' \Sigma^{-1}(x-\mu)),$$

where q is a decreasing function on $[0,\infty)$, and the constant c depends on dimension p and functional form q, but not on μ and Σ. On this family, the readers may be referred to Kelker (1971) and Kariya and Sinha (1989). The pdf $h(x)$ in (4.9) may not admit the existence of mean and covariance matrix. If they exist,

$$(4.10) \quad E(x) = \mu \quad \text{and} \quad \text{Cov}(x) = \lambda \Sigma \quad (\lambda > 0).$$

We call μ location parameter and Σ scale matrix.

Some examples of the distributions whose pdf's are of the form (4.9) are as follows.

(1) Multivariate power exponential distribution;

$$(4.11) \quad h_1(x|\mu, \Sigma) = c|\Sigma|^{-1/2} \exp[-\frac{1}{2}\{(x-\mu)' \Sigma^{-1}(x-\mu)\}^\alpha],$$

where $\alpha > 0$. The pdf with $\alpha = 1/2$ in (4.11) is the pdf of multivariate double exponential distribution whose tail part is thicker than the pdf of $N(\mu, \Sigma)$ with $\alpha = 1$. If $\alpha \to \infty$, it converges to the pdf of multivariate uniform distribution. The shape parameter α is usually unknown. For testing problems on α and some other properties of distribution, see Kuwana and Kariya (1991). Note $c = \alpha \Gamma(p/2)[2^{1/\alpha}\pi]^{-p/2}[\Gamma(p/2\alpha)]^{-1}$. The mean and covariance matrix exist and are given by (4.10) with $\lambda = 2^{1/\alpha}\Gamma((p+2)/\Gamma(p/2\alpha))$.

(2) Multivariate normal mixture distribution;

$$(4.12) \quad h_2(x|\mu, \Sigma) = \alpha_1 q(x|\mu, \sigma_1 \Sigma) + \alpha_2 q(x|\mu, \sigma_2 \Sigma)$$

where $g(x|\mu, \Sigma)$ is the pdf of $N(\mu, \Sigma)$, $\sigma_1, \sigma_2 > 0$, $\alpha_1 + \alpha_2$ $=1$ and $\alpha_1, \alpha_2 > 0$.

(3) Multivariate t distribution with d.f. (degrees of freedom) n

(4.13)
$$h_3(x|\mu, \Sigma) = c|\Sigma|^{-1/2}[1+\frac{1}{n}(x-\mu)' \Sigma^{-1}(x-\mu)]^{-(n+p)/2}.$$

The pdf with $n=1$ is the pdf of multivariate Cauchy distribution in which no mean and hence no covariance matrix exist. If $n \geq 3$, the mean and covariance matrix exist.

Robust classification rule

Suppose x_0 is an observation from either of

(4.14) $\Pi_i : f_i(x) = h(x|\mu_i, \Sigma_i)$ $(i=1,2)$,

where the pdf h is of the form (4.9). In this subsection (μ_i, Σ_i)'s ($i=1, 2$) are assumed to be known. Then by the Lemma the optimal classification rule is to assign x_0 to Π_i according as $x_0 \in A_i$ with

(4.15) $A_1 = \{x : R(x) \geq k\}$, $A_2 = R^p - A_1$.

where $k = a_2/a_1$ and

(4.16) $R(x) = \dfrac{|\Sigma_1|^{-1/2} q((x-\mu_1)' \Sigma_1^{-1}(x-\mu_1))}{|\Sigma_2|^{-1/2} q((x-\mu_2)' \Sigma_2^{-1}(x-\mu_2))}.$

I.e., x_0 is assigned to Π_1 if $R(x_0) \geq k$ and to Π_2 if $R(x_0) < k$. Thus $R(x)$ or $\log R(x)$ is the optimal classification function. In general, $R(x)$ depends on the functional form q of the pdf's. However, in the following case it is independent of q.

A) Case $k^{2/p}|\Sigma_1| = |\Sigma_2|$: Quadratic classification function.
   In this case, as q is decreasing on $[0, \infty)$, $R(x) \geq k$ is equivalent to $D_1(x) > 0$ with

(4.17) $D_1(x) = (x-\mu_2)' \Sigma_2^{-1}(x-\mu_2) - (x-\mu_1)' \Sigma_1^{-1}(x-\mu_1)$, which is surely independent of q.   Also (4.15) is reduced to

(4.18) $A_1 = \{ x : D_1(x) \geq 0 \}$, $A_2 = \{ x : D_1(x) < 0 \}$.

If $a_1 = a_2 = 1/2$, then $k = 1$ and so the condition $|\Sigma_1| = |\Sigma_2|$ leads to (4.17).

B) $k = 1$ and $\Sigma_1 = \Sigma_2$:   Linear classification function.
   This case is a special case of $|\Sigma_1| = |\Sigma_2|$, and $R(x) \geq 1$ is equivalent to $D_2(x) \geq 0$ with

(4.19) $D_2(x) = 2(\mu_1 - \mu_2)' \Sigma^{-1} x - (\mu_1 - \mu_2)' \Sigma^{-1}(\mu_1 + \mu_2)$,

where $\Sigma = \Sigma_1 = \Sigma_2$.   This is equivalent to (3.11) in Section 3.
   As a special case, let us consider the case of (1) Multivariate power exponential distribution (4.11).   Then we assign x_0 to Π_1 if $\log R(x) \geq \log k$, where

(4.20)
   $\log R(x) = d(x : \mu_2, \Sigma_2)^{\alpha} - d(x : \mu_1, \Sigma_1)^{\alpha} - \frac{p}{2} \log |\Sigma_1| / |\Sigma_2|$,

with $d(x : \mu, \Sigma) = (x - \mu)' \Sigma^{-1}(x - \mu)$.   This function depends on α.   If $\alpha = 1$ (normal distribution) or if $k^{2/p} |\Sigma_1| = |\Sigma_2|$,

(4.21) $\log R(x) = D_1(x) - \frac{p}{2} \log |\Sigma_1| / |\Sigma_2|$.

Further if $k = 1$ and $\Sigma_1 = \Sigma_2$, it is reduced to the linear classification $D_2(x)$ in (4.19).

The case where (μ_i, Σ_i)'s are unknown
   When (μ_i, Σ_i)'s are unknown, we need to estimate them based on the samples whose populations are Π_1 and Π_2;

(4.22) x_{ik} $(k = 1, \cdots, n_i)$ iid $\sim$ $\Pi_i : h(x | \mu_i, \Sigma_i)$ $(i = 1, 2)$,

where $\{ x_{1k} \}$ and $\{ x_{2k} \}$ are independent.   We estimate (μ_i, Σ_i) by

$$(4.23) \quad \begin{aligned} \hat{\mu}_i &= \overline{x}_i = \frac{1}{n_i} \Sigma_{k=1}^{n_i} x_{ik}, \\ \hat{\Sigma}_i &\propto \frac{1}{n_i} \Sigma_{k=1}^{n_i} (x_{ik} - \hat{\mu}_i)(x_{ik} - \hat{\mu}_i)' \equiv S_i. \end{aligned}$$

If q is known, $\hat{\Sigma}_i = \lambda S_i$. The classification rule $R(x)$ with estimate $(\hat{\mu}_i, \hat{\Sigma}_i)$ for (μ_i, Σ_i) is denoted by $\hat{R}(x)$.

Even if q is unknown, if the condition in A) is satisfied, $\hat{D}_1(x)$ with the estimates can be used because $D_1(x)$ does not depend on q and so on λ. Hence we test the hypothesis for A);

$$(4.24) \quad H_A : k^{2/p} |\Sigma_1| = |\Sigma_2|.$$

Taking $n = \min(n_1, n_2)$ and forming S_i in (4.24), H_A is rejected if $|u| > c$, where

$$(4.25) \quad u = \sqrt{n} [\log k^{2/p} |S_1| - \log |S_2|]/2p^{1/2}.$$

When n is large, u is asymptotically distributed as $N(0,1)$ (see Siotani, et.al. (1985)). If H_A is not rejected, we may use $\hat{D}_1(x)$ for classification, and further we may test the hypothesis for B);

$$(4.26) \quad H_B : \Sigma_1 = \Sigma_2 \quad \text{with} \quad k=1.$$

The testing procedure for H_B has been given in Section 3. If H_B is not rejected, we may use $\hat{D}_2(x)$ for classification.

Exercises

1. Choose some ancillary variables (such as PER, PBR, etc.), and check the predictive power of the variables through the IC method.
2. Try a classification rule in Section 3 to form a portfolio and check the performance.
3. Study on the proof of Lemma in Section 4.
4. For the pdf (4.11), compute the covariance matrix.

CHAPTER 10

OPTIMAL MTV MARKET PORTFOLIO

1 MTV-MP system

In this chapter we introduce what we call the optimal MTV market portfolio construction system (MTV-MP system below), in which the MTV model is set as a core model for the analysis of price fluctuations and the concept of prediction is centered for the construction of an optimal portfolio. A main idea of this system has been realized in the New Japan Securities MTV Portfolio System.

The flow chart of the MTV-MP system is as follows:

1) To select a target population which consists of 50~70 stocks from the total population such as the Section 1 of the Tokyo Stock Market which lists 1200 stocks approximately. The criteria of the selection are
   a) for the forecasted returns to exceed a certain level and
   b) for the price fluctuations to have high commonalities with the fluctuations of multiple common market factors.
2) To select an investment stance such as a target value of portfolio return.
3) Analysis of price fluctuations, prediction and evaluation of prediction errors via an MTV model.
4) Optimization through controlling the model fit and the prediction error for future portfolio returns.
5) Construction of an optimal portfolio, a modified portfolio, etc.
6) To repeat 4) and 5) for various investment stances.
7) With new data arriving, to construct new optimal portfolios, compare them with the portfolios under monitoring and management, and to rebalance the old portfolios via a rebalance rule if necessary.

It is a feature of the MTV-MP system that future values of

returns are predicted via an MTV model. Of course, in the part of
selecting a target population the construction of a portfolio, the
accuracy and stability of the predictive power of the MTV model to
be formulated is the most important element to be taken into
account. For this purpose multiple common market variation components that drive major parts of the market and move rather systematically and stably are identified and then a group of stocks
that move together with these components is chosen for modelling by
a random grouping method. Also in the analysis of price variations
via an MTV model a great deal of statistical considerations and
diagonastics are made for the statistical stability and predictive
power of the model. Further in the part of optimization the error
rate of the model for a given investment stance is minimized for
the stability and accuracy again. When new data arrives, the model
is reestimated based on the new common market components, the predicted values are revised, and the future performances of the old
and new optimal portfolios are forecasted, compared and evaluated.

2   Extraction of common market components and selection of a target population

The problem of selecting a target population is relative to the
object of the analysis. For a predictive purpose we cannot take a
long sample period (see Chapter 3) and so it will be necessary to
make the number of assets for each analysis relatively small, obtain a stable result of the analysis and secure the predictability.
This is a problem in selecting a target population from a universe
of the assets. For example, there are approximately 1,500 stocks
listed in the Tokyo Stock Market and to treat these stocks simultaneously, we will need at least 3,000 data for stable modelling,
which requires us at least 12 years daily data, 60 years weekly
data, etc. The choice of the length of one period basically
depends on investment stance and the cost of carrying out rebal-

ances and managing portfolios. Hence a question here is whether 12
years is a proper length to construct a model for predicting daily
fluctuations of prices.   In our approach, we attempt to extract
common market variance components, whose number and whose processes
are very likely to be changing over 12 years.   The model we need is
a model reflecting recent fluctuations so that it will have a predictive power for near future.

In practice, it is also important to transform data in association with the object of modelling. A convenient form of price data
for portfolio analysis, as has been discussed, will be

$$(2.1) \quad x_{it} = S_{it} / S_{iT} \quad (i=1, \cdots, N: \ t=1, \cdots, T),$$

where T is the (present) time of investment.   Or one may use
standardized prices for x_{it}'s.

Stage 1   Screening by forecasted returns

(1) Devide N stocks randomly into M groups each of which consists
of e_m stocks ($50 \leq e_m \leq 70$, say), so that

$$e_1 + \cdots + e_M = N.$$

(2) Repeat (1) K times so that we can obtain K blocks of M
groups for the same N stocks where K is specified in advance or
$K=[N/100]+1$ where $[a]$ denotes the greatest integer not larger
than a.   To each group m ($m=1, \cdots, M$) of the k-th block ($k=1,$
$\cdots, K$), apply an MTV model to predict the $(T+j)$-th value of x_{it}
($j=1, \cdots, J$) in (2.1). J depends on the unit of one period (daily, weekly or monthly).   Let

$$(2.2) \quad x_{it}(k, m) \quad \text{and} \quad \hat{x}_{it}(k, m)$$

denote the value and the predicted value of x_{it} in the m-th group
of the k-th block respectively.   Note that we may use a nonlinear
MTV model in prediction.

(3) Let ν_t be a weight for the t-th predicted value and let the weighted average of predicted values $\hat{x}_{it}(k, m)$ in (2.2) over $t = T+1, \cdots, T+ J$ be

$$\bar{\hat{x}}_i(k, m) = \Sigma_{t=T+1}^{T+J} \nu_t \, \hat{x}_{it}(k, m)/\Sigma_{t=T+1}^{T+J} \nu_t.$$

Collect the stocks for which $\bar{\hat{x}}_i(k, m)$ is greater than a specifi-ed level (say, an expected interest rate in the J periods) and let $q(k, m)$ denote the number of those stocks. In this stage, various restrictions on $\bar{\hat{x}}_i(k, m)$ can be imposed through ν_t's and an investment stance. Collecting the stocks meeting with these demands from each group, there are $q(k) = \Sigma_{m=1}^{M} q(k, m)$ stocks collected in the k-th block, whose set of collected stocks is denoted by $B(k)$.

(4) Repeating (3) for each block $k=1, \cdots, K$, we obtain $B(1), \cdots$, $B(K)$. In general, there are many multiplicities of stocks in these blocks $B(k)$'s. Hence the number of stocks contained in the union $\cup_{k=1}^{K} B(k)$ is much smaller than $q(k) \times K$ and is denoted by $N(1)$. If necessary, we can delete some stocks which are consider-ed improper on the basis of some other screening criteria.

(5) If the number $N(1)$ of stocks selected in (4) is greater than a pre-specified number L, repeat (1)~(4) for the $N(1)$ stocks. In each repetition, parameters in the procedures (1)~(4) may be changed. And if $N(\ell) < L$ is attained at the ℓ-th repetition, go to (6) in the second stage.

Stage 2   Screening by commonalities with market

(6) The $N(\ell) \equiv N_0$ stocks are randomly divided into M_0 groups of e_m stocks in the same way as (1). If $M_0 = 1$, go to (10).

(7) Apply the MTV principal component analysis based on the covari-ance matrix of x_{it}'s to the m-th group, let n_m denote the smal-lest number of the principal components whose cumulative relative contribution is more than a specified value and let $f_{jt}(m)$ ($j=1$,

$\cdots$, n_m) denote the n_m principal components. Then compute the commonalities of the i-th stock in the m-th group with these n_m components, which are denoted by

$$c(m, \ i) \quad (\ i=1, \cdots, \ e_m).$$

Ordering these e_m commonalities for the m-th group in their sizes, select the stocks whose commonalities are more than a specified value and let us call the selected stocks $i=1, \cdots, \ p_m$ anew. This procedure is repeated for $m=1, \cdots, M_0$.

(8) Combine the selected principal components in each group into one vector for $m=1, \cdots, M_0$ as

$$f_t \equiv (\ f_{1\,t}(1), \cdots, \ f_{n_1\,t}(1): \ \cdots \ ; \ f_{1\,t}(M_0), \cdots, \ f_{n_{M_0}}(M_0))' : n \times 1,$$
$$n_1 + n_2 + \ \cdots \ + n_{M_0} = n.$$

Then apply a covariance principal component analysis to $\{\ f_t: \ t=1, \cdots, \ T\}$ and pick the main principal components whose cumulative relative contribution is more than a specified level, which are denoted by $(\ h_{1\,t}, \cdots, \ h_{c\,t})$.

(9) Regress the p_m stock price ratios $x_{i\,t}$'s on the common principal components $(\ h_{1\,t}, \cdots, \ h_{c\,t})$ one by one to get p_m regression equations and repeat it for $m=1, \cdots, M$ so that $p_1 + \cdots + p_{M_0}$ regression equations are obtained. In this procedure, a best regression equation for each $x_{i\,t}$ must be selected based on a certain criterion for selecting regression variables. Then select the Q stocks whose regression equations have larger coefficients of determination. In this stage some additional conditions will be imposed associated with investment stances.

(10) The Q stocks thus selected form a target population.

In the first stage a subgroup of N_0 stocks whose predicted returns are relatively high over J periods is chosen. In the second stage, c common market variation components $(\ h_{1\,t}, \cdots, h_{c\,t})$ which jointly express the market streams are extracted from the subgroup of the N_0 stocks and then the stocks whose returns $\{x_{it}\}$ are mostly explained by the common market components are selected. In other words, the second stage aims to extract *stable* common market components and select some stock returns which have a strong correlation with the common components in the market. This will secure a stable predictability.

3 Optimal MTV portfolio

Once a portfolio population is selected, we predict future returns and risks of the assets in the population via MTV model and form an optimal portfolio based on the predictive retruns and risks. Let $x_{it} = S_{it}/S_{iT}$ as before. Note that $x_{it} - 1$ for $t > T$ is future return relative to the price S_{iT} at T, the time point of investment. Let $\hat{x}_{it}$ denote the predicted value of x_{it} for $t > T$ and the estimated (model) value of x_{it} for $t \leq T$.

In MTV model, each x_{it} is expressed as

$$(3.1) \quad x_{it} = \alpha_{i0} + \alpha_{i1} f_{1t} + \cdots + \alpha_{iq} f_{qt} + \varepsilon_{it},$$

which is modelled based on data $\{x_{it} : i = 1, \cdots, p, \quad t = 1, \cdots, T\}$ as

$$(3.2) \quad x_{it} = \hat{x}_{it} + e_{it}$$
$$= \hat{\alpha}_{i0} + \hat{\alpha}_{i1} \hat{f}_{1t} + \cdots + \hat{\alpha}_{iq} \hat{f}_{tq} + e_{it}. \quad (t = 1, \cdots, T)$$

Here $\hat{f}_{jt}$ is the predicted value of variance component f_{jt} at time $t - 1$ via a time series model such as an AR model, a GARCH model, etc. (see Chapter 5), and e_{it} is the residual which supposedly carries no significant information on the time series process of $\{x_{it} : i = 1, \cdots, p\}$ though e_{it} may be cross-sectionally correlated with e_{jt}. It is noted that by the form of $x_{it} = S_{it}/S_{iT}$ for price variables, the variance of ε_{it} and the covariance of ε_{it} and ε_{jt} may depend on time t. But in the sequal, without modelling then, we simply assume that they are constant;

(3.3) $\sigma_{ij} = \text{Cov}(\varepsilon_{it}, \varepsilon_{jt})$ for $T - t_0 \leq t \leq T + J$,

where J is a prediction period. Hence though $x_{iT} \equiv 1$, x_{iT} contains error ε_{iT}. The constant variances and covariances in (3.3)
are estimated by

(3.4) $\hat{\sigma}_{ijT} = \dfrac{1}{T - t_0} \Sigma_{t=T-t_0}^{T} e_{it} e_{jt}$ $(i, j = 1, \cdots, p)$.

As we cannot control future ε_{it}'s $(t > T)$, we simply predict the
future value of x_{iT+k} by $\hat{x}_{iT+k}$ in (3.2) as

(3.5) $\hat{x}_{iT+k} = \hat{\alpha}_{i0} + \hat{\alpha}_{i1} \hat{f}_{1T+k} + \cdots + \hat{\alpha}_{iq} \hat{f}_{qT+k}$ $(k = 1, \cdots, J)$.

The $\hat{x}_{iT+k}$ in (3.5) is regarded as the conditional mean of x_{iT+k}
and its estimated variance of prediction error is given by

(3.6) $\hat{\sigma}_{ijT+k} = \Sigma_{m=1}^{q} \hat{\alpha}_{im} \hat{\alpha}_{jm} \hat{\omega}_{mT+k} + \hat{\sigma}_{ijT}$,

where $\hat{\omega}_{mT+k}$ is an estimate of the variance of prediction error of
$\hat{f}_{mT+k}$ for f_{mT+k} (see below). If $\{f_{mt}\}$'s are modelled by AR
models, ω_{mT+k} is of the form $\gamma_{T+k}(\phi)\omega_m$ where $\gamma_{T+k}(\cdot)$ is a
function of the vector ϕ of AR coefficients and ω_m is the vari-
ance of an error term in the AR model (see below). In (3.6), the
smaller the error variance $\hat{\sigma}_{iiT}$ or the prediction period k is,
the smaller the (estimated) variance of prediction error is. However, in general the error variance $\hat{\sigma}_{iiT}$ defined by (3.4) is not
small for individual returns and hence the prediction error variance will be considerably large. This implies that the prediction
of x_{iT+k} by $\hat{x}_{iT+k}$ is not sufficiently reliable as it stands. If
x_{iT+k} happens to be distributed as normal with mean $\hat{x}_{iT+k}$ and
variance $\hat{\sigma}_{iiT+k}$, then the probability that x_{iT+k} falls in the
interval $[\hat{x}_{iT+k} - 2\hat{\sigma}_{iiT+k}, \hat{x}_{iT+k} + 2\hat{\sigma}_{iiT+k}]$ is approximately
0.95. But this interval is usually relatively large. What will
make smaller the prediction error variance is the portfolio effect
which consists of diversification effect and Markowitz effect

(Chapter 6).

Optimal prediction portfolio

As is discussed in Chapter 6, future portfolio returns when invested at time T are expressed as

$$(3.7)\quad X_{PT+k} = X_{PT+k}(d) = d_{1T}X_{1T+k} + \cdots + d_{pT}X_{pT+k},$$

which is predicted by

$$(3.8)\quad \hat{X}_{PT+k} = \hat{X}_{PT+k}(d) = d_{1T}\hat{X}_{1T+k} + \cdots + d_{pT}\hat{X}_{pT+k},$$

where d_{iT}'s are portfolio value ratios and $d=(d_{1T}, \cdots, d_{pT})'$. By (3.2) the difference between (3.7) and (3.8) is expressed as

$$(3.9)\quad X_{PT+k} - \hat{X}_{PT+k} = \Sigma_{i=1}^{P} d_{iT}(X_{iT+k} - \hat{X}_{iT+k}).$$

But each X_{iT+k} follows the model (3.1) with $t = T+k$. Hence as will be given in the next section, the prediction error variance of the portfolio return at $T+k$ is estimated by

$$(3.10)\quad \hat{\sigma}_{T+k}^{2}(d) = \Sigma_{m=1}^{q}(d'\hat{\alpha}_m)^2\hat{\omega}_{mT+k} + d'\hat{\Sigma}d$$

$$= d'[\hat{A}_q\hat{\Omega}_{T+k}\hat{A}_q' + \hat{\Sigma}]d$$

where $\hat{\alpha}_m = (\hat{\alpha}_{1m}, \cdots, \hat{\alpha}_{pm})'$, $\hat{\Sigma} = (\hat{\sigma}_{ijT})$, $\hat{A}_q = [\hat{\alpha}_1, \cdots, \hat{\alpha}_q]$ and $\hat{\Omega}_{T+k} = \mathrm{diag}\{\hat{\omega}_{1T+k}, \cdots, \hat{\omega}_{qT+k}\}$. This expression of the risk of the portfolio return at $T+k$ is not the same as the usual historical risk (variance) given as $d'\hat{\Sigma}d$ and it varies with $k=1, \cdots, J$.

Now that the predictive return $\hat{X}_{PT+k}(d)$ and predictive risk $\hat{\sigma}_{t+k}^{2}(d)$ of a portfolio d is obtained, an optimal portfolio is also obtained in such a manner as described in Chapter 6 so long as an objective function is given in terms of $\hat{X}_{PT+k}(d)$ and $\hat{\sigma}_{T+k}(d)$. Typically a predictive Markowitz type optimal portfolio is derived by maximizing

(3.11) $\quad \Phi(d) = \lambda(\hat{x}_{PT+k}(d))^2 - \hat{\sigma}_{T+k}^2(d)$

with respect to d for given k under

(3.12) $\quad 0 \leq d_{iT} \leq D \quad$ and $\quad \Sigma_{i=1}^{P} d_{iT} = 1.$

We may impose $\hat{x}_{PT+k}(d) \geq r_0$ on the maximization.

4 Prediction errors in MTV model

When variables follow an MTV model, x_{iT+m} and its predicted value are given by

(4.1)
$$x_{iT+m} = \mu + \alpha_{i1} f_{1T+m} + \cdots + \alpha_{iq} f_{qT+m} + \varepsilon_{iT+m}$$
$$\hat{x}_{iT+m} = \mu + \alpha_{i1} \hat{f}_{1T+m} + \cdots + \alpha_{iq} \hat{f}_{qT+m}$$

provided parameters are known. Hence the predicted error is

(4.2) $\quad e_{iT+m} = x_{iT+m} - \hat{x}_{iT+m} = \Sigma_{j=1}^{q} \alpha_{ij} \delta_{jT+m} + \varepsilon_{iT+m} \quad$ with

$$\delta_{jT+m} = f_{jT+m} - \hat{f}_{jT+m}.$$

Hence setting $\omega_{jT+m} = \text{Var}(\delta_{jT+m})$ and $\sigma_{ij} = \text{Cov}(\varepsilon_{it}, \varepsilon_{jt})$, we obtain

(4.3)
$$\text{Var}(e_{iT+m}) = \Sigma_{j=1}^{q} \alpha_{ij}^2 \omega_{jT+m} + \sigma_{ii},$$
$$\text{Cov}(e_{iT+m}, e_{kT+m}) = \Sigma_{j=1}^{q} \alpha_{ij} \alpha_{kj} \omega_{jT+m} + \sigma_{ik}.$$

For a portfolio $x_{Pt} = \Sigma_{i=1}^{P} d_i x_{it}$, $e_{PT+m} = x_{PT+m} - \hat{x}_{PT+m}$ and hence its error variance is evaluated as

(4.4) $\quad \text{Var}(e_{PT+m}) = \Sigma_i \Sigma_k d_i d_k \text{Cov}(e_{iT+m}, e_{kT+m})$
$$= \Sigma_{j=1}^{q} (\Sigma_{i=1}^{P} d_i \alpha_{ij})^2 \omega_{jT+m} + \Sigma_{i=1}^{P} \Sigma_{k=1}^{P} d_i d_k \sigma_{ik}.$$

In applications, we may use $\hat{\alpha}_{ij}$ for α_{ij} and $\hat{\sigma}_{ij}$ for σ_{ij} with

$$\hat{\sigma}_{ij} = \Sigma_{t=T-c}^{T} e_{it} e_{jt} / (T-c) \quad \text{where} \quad e_{it} = x_{it} - \hat{x}_{it}.$$

On the other hand, an evaluation of ω_{jT+m} depends on the specification of the model of $\{f_{jt}\}$. If $\{f_{jt}\}$ follows an AR(a) model

$$f_{jt} = \phi_{j1} f_{jt-1} + \cdots + \phi_{ja} f_{jt-a} + \eta_{jt},$$

then

$$\delta_{jT+m} = f_{jT+m} - \hat{f}_{jT+m} = \Sigma_{k=1}^{a} \phi_{jk} \tilde{\eta}_{jT+m-k} + \eta_{jT+m}$$

with $\tilde{\eta}_{jT+m-k} = \eta_{jT+m-k}$ if $m > k$ and $\tilde{\eta}_{jT+m-k} = 0$ if $m \leq k$.
Hence

$$\omega_{jT+1} = E(\delta_{jT+1}^2) = \mathrm{Var}(\eta_{jT+1}) \equiv \Delta_j,$$
$$\omega_{jT+2} = \mathrm{Var}(\eta_{jT+2}) + \phi_{j1}^2 \mathrm{Var}(\eta_{jT+1}) = (1 + \phi_{j1}^2) \Delta_j,$$
$$\omega_{jT+3} = \mathrm{Var}[\eta_{jT+3} + \phi_{j1}(\eta_{jT+2} + \phi_{j1} \eta_{jT+1}) + \phi_{j2} \eta_{jT+1}],$$
$$= [1 + \phi_{j1}^2 + (\phi_{j1}^2 + \phi_{j2})^2] \Delta_j$$

etc. Here the estimates $\hat{\phi}_{jk}$'s and $\hat{\Delta}_j$ are substituted for ϕ_{jk}'s and Δ_j. With these evaluations, an estimated standard deviation of prediction error for x_{PT+m} is obtained by $[\mathrm{Var}(e_{PT+m})]^{1/2}$, and a confidence interval for x_{PT+m} is given by

$$\hat{x}_{PT+m} - k[\mathrm{Var}(e_{PT+m})]^{1/2} \leq x_{PT+m}$$
$$\leq \hat{x}_{PT+m} + k[\mathrm{Var}(e_{PT+m})]^{1/2}.$$

It should be noted that if $x_{it} = S_{it}/S_{iT}$, the heteroscedasticity of $\{x_{it}\}$ and hence $\{f_{jt}\}$ needs to be taken into account in applications.

Exercises

1. Discuss about the random grouping method in the MTV-MP system.
2. Compare the optimal predictive portfolio which minimizes (3.11) with Markowitz type portfolio.
3. Show (4.4).

CHAPTER 11

INDEX PORTFOLIO AND CANONICAL CORRELATION PORTFOLIO

1 Basic concept of index portfolio

As is well known, an index-tracking portfolio or simply index portfolio is a portfolio which performs in terms of returns as well as a pre-chosen market index or a pre-chosen reference variable such as Nikkei 225, S&P 500, etc. To be precise, let T be the time point at which an index portfolio is made, and let S_{0t} be a reference variable for an index portfolio to follow, and S_{it} the price of the i-th asset (stock) at t ($i=1,\cdots, p$), where $\{ t=1, \cdots, T\}$ is the sample period. Then, as has been discussed in Chapter 6, the return process of a portfolio made at T is expressed as $\{\widetilde{x}_{Pt}\}$ with

$$(1.1) \quad \widetilde{x}_{Pt} = d_{1T}\widetilde{x}_{1t}+ \cdots + d_{pT}\widetilde{x}_{pt} \quad (t>T)$$

where $\widetilde{x}_{it}=S_{it}/S_{iT}$, and d_{iT}'s are portfolio value ratios. A portfolio is defined to be an index portfolio for a given reference variable $\{ S_{0t}\}$ if the return process $\{\widetilde{x}_{Pt}\}$ of the portfolio is close to the return process $\{\widetilde{x}_{0t}\}$ with $\widetilde{x}_{0t}=S_{0t}/S_{0T}$, or equivalently if the variance of $\widetilde{x}_{Pt}-\widetilde{x}_{0t}$ is close to zero;

$$(1.2) \quad \text{Var}(\widetilde{x}_{Pt}-\widetilde{x}_{0t}) \approx 0 \quad \text{for} \quad t>T.$$

In our discussion below, the reference variable can be anything although it may be difficult to make an index portfolio of stocks for such a reference variable as exchange rate. To make a "long living" index portfolio without rebalancing, it is important

(1) to select a proper set (portfolio population) of assets ($i=1, \cdots, p$) in which each $\{\widetilde{x}_{it}\}$ is of similar time series variational features as $\{\widetilde{x}_{0t}\}$, and

(2) to make an optimal index portfolio from the selected set of assets which at least minimizes (1.2) over the sample period

$t=1, \cdots, T$.

When the reference variable is Nikkei 225, we can make exactly the same portfolio as Nikkei 225 which makes the variance in (1.2) zero for all t if we buy all the stocks in Nikkei 225 equally. However, one may wish to make an index portfolio for Nikkei 225 with a smaller number of stocks because of a limited fund, or with a better performance in the sense that either

$$(1.3) \quad \text{Var}(\widetilde{x}_{Pt} - \alpha_t - \widetilde{x}_{0t}) \approx 0 \quad (t>T) \quad \text{or}$$

$$(1.4) \quad \widetilde{x}_{Pt} > \widetilde{x}_{0t} \qquad\qquad (t>T).$$

If $\alpha_t = (t-T)\alpha/D$ where $D=365$ for daily data or $D=365/7$ for weekly data, etc., a portfolio with (1.3) is called an index-plus-α portfolio where $100\alpha\%$ is annual rate of excess return over the reference variable. On the other hand, a portfolio with (1.4) is attractive because it is better than the reference variable for all $t>T$. A more appealing portfolio will be

$$(1.5) \quad \widetilde{x}_{Pt} > \widetilde{x}_{0t} + \alpha_t \quad \text{for} \quad t>T,$$

though it may be difficult to make such a portfolio. An idea to make a portfolio with (1.3) is that we choose as a reference variable,

$$(1.6) \quad S_{0t}^{\alpha} = (1+\alpha/D)^t S_{0t} \qquad (t=1, \cdots, T),$$

which grows $\alpha\%$ more than S_0, and make an index portfolio for $\{S_{0t}^{\alpha}\}$. Hence if a method to make an index portfolio for a given reference variable is available, we may be able to make a portfolio satisfying (1.3).

In the literature, probably because of an influence of the MPT, one-period return $x_{it} = \log S_{it} - \log S_{it-1}$ ($i=0,1,\cdots, p$) is customarily used for analysis, in which case the performance of an index portfolio should be evaluated in terms of $\widetilde{x}_{Pt}^* = \sum_{i=1}^{p} d_{iT} \widetilde{x}_{it}^*$

where

(1.7) $\widetilde{x}_{it}^* = x_{it} + x_{it-1} + \cdots + x_{iT+1}$ $(t > T)$.

But the problem (1) of selecting a portfolio population is not well discussed in the literature probably because the time series structures of x_{it}'s are rather weak, as has been observed in Chapter 2. When a market index such as S&P500 is taken as a reference variable, random sampling or stratified sampling is often used or sometimes recommended for selecting a population.   However, such a procedure will fail to give an efficient or long-living index portfolio because price processes of different time series structure destabilize the portfolio and expand the tracking errors soon after sample period.   The problem (2) of forming an optimal portfolio is treated traditionally based on a multifactor model for $\{x_{it}\}$.   As such a multifactor model, there are many versions with respect to choices of factors (see Chapters 7 and 8).   In a factor analysis model, factors are implicit, while in Rosenberg's model factors called descriptors are various variables representing fundamental indicators of each firm.

2 MTV index portfolio
Selection of a portfolio population
As was discussed in Section 1, it is quite important to select a proper set of stocks (or assets) whose return processes $\{\widetilde{x}_{it}\}$ are of similar time series features as the return process $\{\widetilde{x}_{0t}\}$ of a given reference variable.   We consider the following approaches to this problem.

A MTV model appproach.

B Cluster analysis approach.

In the MTV model approach, the variations of returns of $\widetilde{x}_{it}$'s are modelled via an MTV model and stocks whose return processes are close to $\{\widetilde{x}_{0t}\}$ are chosen.   We suppose there are N stocks in the universe and $T \geqq 100$.

(1) Decompose randomly N stocks into M groups each of which consists of N_k stocks ($50 \leqq N_k \leqq 70$) (see Chapter 10), so that

$$N_1 + \cdots + N_M = N.$$

(2) Add the reference variable ($i = 0$) to each group and obtain a factor-loading expression of MTV model (see Chapter 5):

(2.1) $\widetilde{x}_{it} = \beta_{i0} + \beta_{i1} g_{1t} + \cdots + \beta_{iq} g_{qt} + \varepsilon_{it}$ ($i = 0, 1, \cdots, N_k$).

Here the factors are renumbered in the descending order of β_{0j}^2 ($j = 1, 2, \cdots, q$) and q is chosen in order that $\beta_{01}^2 + \cdots + \beta_{0q}^2$ be greater than a certain level.

(3) Compute the distance

(2.3) $d(i, 0) = \omega_1 (\beta_{i1} - \beta_{01})^2 + \cdots + \omega_q (\beta_{iq} - \beta_{0q})^2$

with suitable weights ω_i's and choose stock i's whose distances are less than a specified number. This distance measures a similarity or closeness of the time series structures of i and 0.

(4) Repeat the procedures (2) and (3) to get $50 \sim 70$ stocks for a portfolio population.

In the cluster analysis approach, a similarity of two return processes is directly defined by a quantitative measure. A typical measure is the Euclidean distance $D_1(i, 0)$ between the i-th process and the 0-th process;

(2.3) $D_1(i, 0) = \Sigma_{t=1}^{T} (\widetilde{x}_{it} - \widetilde{x}_{0t})^2.$

There are many other measures for similarity. Whatever it may be, once it is defined with certain legitimacy, stocks whose distances are smaller are chosen for a portfolio population. The details are given in Section 3.

Construction of MTV index portfolio

For a given portfolio population ($i=1, \cdots, p$), we add the reference variable and obtain a factor-loading expression of MTV model;

$$(2.4) \quad \widetilde{x}_{it} = \gamma_{i0} + \gamma_{i1} g_{1t} + \cdots + \gamma_{iq} g_{qt} + \varepsilon_{it}.$$

Then the return process of a portfolio is expressed as

$$(2.5) \quad \widetilde{x}_{Pt} = \gamma_{P0} + \gamma_{P1} g_{1t} + \cdots + \gamma_{Pq} g_{qt} + \varepsilon_{Pt} \quad \text{with}$$

$$(2.6) \quad \widetilde{x}_{Pt} = \Sigma_{i=1}^{P} d_i \widetilde{x}_{it}, \quad \gamma_{Pj} = \Sigma_{i=1}^{P} d_i \gamma_{ij} \quad \text{and}$$
$$\varepsilon_{Pt} = \Sigma_{i=1}^{P} d_i \varepsilon_{it} \quad (j=0,1,\cdots, q).$$

Now to get an optimal portfolio, the coefficients γ_{Pj}'s of this portfolio return are set equal to those of $\widetilde{x}_{0t}$;

$$(2.7) \quad \gamma_{Pj} = \gamma_{0j} \quad (j=0,1,\cdots, q),$$

and minimizing

$$(2.8) \quad \Sigma_{t=1}^{T} (\widetilde{x}_{Pt} - \widetilde{x}_{0t})^2.$$

with respect to d_i's under (2.7) probably with $0 \leq d_i \leq a$ yields an optimal index portfolio.

3 Cluster analysis for grouping

In selecting a portfolio population for forming an index portfolio, we may use cluster analysis, a multivariate method which groups given individual objects (assets) on the basis of a measure of similarity or dissimilarity in terms of a specific feature of objects. The measure of similarity or dissimilarity is often represented by a distance between two objects or two groups. For example, in Section 2 we defined a measure of distance $d(i, 0)$ between the 0-th asset and the i-th asset by using the coefficients of MTV model, which can be directly generalized into a measure of distance $d(i, j)$ between the i-th and j-th assets as

$$(3.1) \quad \widetilde{d}(i, j) = \omega_1(\beta_{i1} - \beta_{j1})^2 + \cdots + \omega_q(\beta_{iq} - \beta_{jq})^2.$$

If the distance is smaller, then the movements of returns of the i-th and j-th assets are similar because it implies a similarity of the coefficients. A detailed procedure of cluster analysis is as follows.

(1) Now suppose there are N objects (assets) to be grouped, say $i = 1, \cdots, N$, which are denoted by

$$G_i(1) = \{i\} \quad (i = 1, \cdots, N),$$

meaning that there are N groups each of which consists of only one object at the outset. It is also assumed that a distance $d(G_i(1), G_j(1)) \equiv d(\{i\}, \{j\})$ between two groups is defined over $G_i(1)$'s. An example of such a distance is $\widetilde{d}(i, j)$ in (3.1). In the following procedure, the number of groups is reduced one by one at each step by combining two closest groups into one group.

(2) (2nd step) To choose one pair out of $G_i(1)$'s and make a union, compute the distance $d(G_i(1), G_j(1))$ for all pairs $(G_i(1), G_j(1))$'s $(i \neq j)$ and choose the pair which minimizes the distance. Without loss of generality, let the pair be $(G_{N-1}(1), G_N(1))$ and form $G_{N-1}(2) = G_{N-1}(1) \cup G_N(1) = \{N-1, N\}$. Also let $G_i(2) = G_i(1)$ for $i = 1, \cdots, N-2$. Hence there are $N-1$ groups in the second step.

(3) (3rd step) For given $G_1(2), \cdots, G_{N-2}(2), G_{N-1}(2)$, we compute the distances of all the pairs $(G_i(2), G_j(2))$ $(i \neq j)$ to form a second union out of the groups. For $i, j = 1, \cdots, N-2$, the distance of a pair is of course given by

$$d(G_i(2), G_j(2)) = d(G_i(1), G_j(1)).$$

But for pair $(G_i(2), G_{N-1}(2))$, we need to define a new distance. When a distance $\delta(\{i\}, \{j\})$ is defined over individual objects, one of the following distances are often used as a distance between two groups $H_i=\{ i_1, \cdots, i_m\}$ and $H_j=\{ j_1, \cdots, j_n\}$;

  i) $\delta_1(H_i, H_j) = \min\{ \delta(\{ k\},\{ \ell\}) \mid k \in H_i, \ell \in H_j\}.$

  ii) $\delta_2(H_i, H_j) = \max\{ \delta(\{ k\},\{ \ell\}) \mid k \in H_i, \ell \in H_j\}.$

Hence if we choose the least distance i) between two groups, for pair $(G_i(2), G_{N-1}(2))$ we compute

$$d_1(G_i(2), G_{N-1}(2)) = \min\{ d(\{ i\},\{ \ell\}) \mid \ell \in G_{N-1}(2)\},$$

and since $d(G_i(2), G_j(2))= d_1(G_i(2), G_j(2))$ for $i, j=1, \cdots,$ N-2, we make a union of the two groups which give the smallest distance $d_1(G_i(2), G_j(2))$ in pair (i, j) where $i, j=1, \cdots, N$-1 and $i \neq j$. If the pair (i, j) minimizing $d_1(\cdot, \cdot)$ is (i, j) with $i, j \leq N$-2, then without loss of generality let it be $(G_{N-3}(2),$ $G_{N-2}(2))$ and let $G_{N-3}(3)= G_{N-3}(2) \cup G_{N-2}(2)=\{ N$-3, N-2$\}$, $G_{N-2}(3)= G_{N-1}(2)=\{ N$-1, $N\}$ and $G_i(3)= G_i(2)= G_i(1)$ for i $=1, \cdots, N$-4. If the pair (i, j) minimizing $d_1(\cdot, \cdot)$ is $(i, N-1)$ with $i \leq N-2$, without loss of generality let i be $N-2$ and let $G_{N-2}(3)= G_{N-2}(2) \cup G_{N-1}(2)=\{ N-2, N-1, N\}$ and $G_i(3)= G_i(2)$ for $i=1, \cdots, N$-3. Hence there are $N-2$ groups at the third step.

    Repeating this procedure, at the k-th step we have $N- k+1$ groups and hence at the N-th step all the objects are combined into one group. At each step, the minimum distance over pairs of groups is computed, which is used to evaluate how close two groups combined at each step are.

    This grouping (clustering) greatly depends on the choice of the distances at the first and second steps. In many financial analyses, no model is used for generation of prices or returns and some distances are defined over return (or price) series $\{ x_{it}\}$'s without paying any attention to the time series structure of $\{ x_{it}\}$'s. Typically the following distances are often adopted:

(1) Euclidean distance;

$$d(\{ i\},\{ j\}) = [\Sigma_{t=1}^{T}(x_{it}- x_{jt})^2]^{1/2}$$

(2) Standardized Euclidean distance; with $z_{it}=(x_{it}-\overline{x}_t)/\sqrt{s_{tt}}$

$$d(\{i\},\{j\}) = [\Sigma_{t=1}^{T}(z_{it}-z_{jt})^2]^{1/2}$$

where $\overline{x}_t=\Sigma_{i=1}^{N}x_{it}/N$ and $s_{ts}=\Sigma_{i=1}^{N}(x_{it}-\overline{x}_t)(x_{is}-\overline{x}_s)/N.$
(3) Mahalanobis distance;

$$d(\{i\},\{j\}) = [(x_i-x_j)'\,S^{-1}(x_i-x_j)]^{1/2}$$

where $x_i=(x_{i1},\cdots,x_{iT})'$ and $S=(s_{ts}).$

In the literature there are a great deal of distances proposed. Most of them are of cross-sectional and descriptive feature. In fact, cluster analysis itself is a descriptive method and a blind application of the method to a time series data with a wrong distance is likely to yield a meaningless result. For return series $\{x_{it}\}$, the distance $\widetilde{d}(i,j)$ in (3.1) is a promising one.

4   Index plus α premium minus β risk historical portfolio

For a given reference variable (S_0), we consider how to construct a portfolio whose historical average of one-period returns and historical risk (standard deviation of returns) over a past sample period are respectively $\alpha\%$ larger and $\beta\%$ smaller than those of the reference variable annually. We call such a portfolio an $\alpha-\beta$ portfolio in the sequal. This construction method does not make use of the variational structures of stocks or assets and hence it is required to check the validity or effectiveness for extrapolation by simulations. But the variational structure of stocks may not be thought to change rapidly though it may change gradually. Hence an $\alpha-\beta$ portfolio may perform effectively for near future period when it is constructed based on a short sample period. Consequently it is expected that an $\alpha-\beta$ portfolio combined with a rebalance rule will have a performance.

Figure 4-1 plots means and standard deviations of daily returns

of stocks contained in Nikkei 225 and those of Nikkei 225 itself,
from which it is observed that

(1) there is a weak relation of high return-high risk,

(2) the standard deviation of the Index is significantly smaller
    than those of individual stocks, showing that the Index forms a
    risk-diversifying portfolio, and

(3) the mean return of the Index is an average of individual
    returns.

Fig. 4-1 Means and standard deviations of daily returns

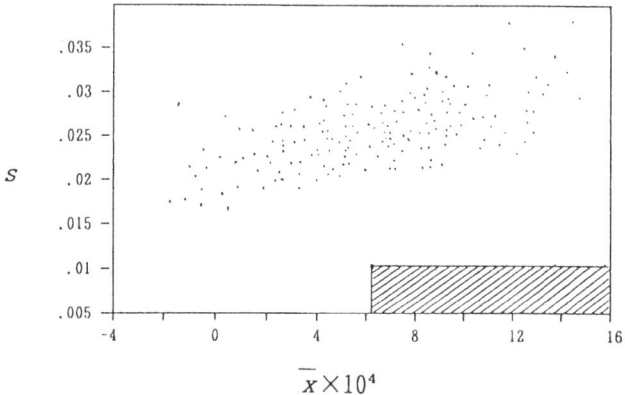

$$\overline{x} \times 10^4$$

The purpose of this section is to construct a portfolio whose
averaged return and standard deviation of returns (risk) sit in
the shadow area in Figure 4-1.   Such a portfolio thus obtained is
at least historically better than the Index in the sense that it
has a larger return and a smaller risk on the average over the
chosen period.   In this example, the sample period is 5 years, but
for a practical use of such a portfolio, it will be better to take
a shorter sample period from the present time so that the historical (past) performance observed in the sample period will continue
in the near future out of sample.

Index $\alpha - \beta$ portfolio

    Let S_{it}'s $(i=1, \cdots, p)$ be prices of stocks in the portfolio

population selected in advance, which may be the set of the stocks
adopted in the Index.   Let $x_{it} = \log S_{it} - \log S_{it-1}$ be the one-
period return of the i-th stock from $t-1$ to t.   Let c_i be the
portfolio unit ratio of the i-th stock, and let

(4.1)
$$Q_t(c) = \Sigma_{i=1}^{p} c_i S_{it} \quad \text{and} \quad x_t(c) = \log Q_t(c) - \log Q_{t-1}(c)$$

be the price and one-period return of the portfolio $c = (c_1, \cdots,$
$c_p)'$ respectively.   Here c belongs to the simplex

(4.2) $C = \{ c \in R^p |\ c_i \geq 0, \Sigma_{i=1}^{p} c_i = 1\}.$

The average and standard deviation of $x_t(c)$'s over a period $t=1$,
$\cdots$, T are respectively

(4.3) $\overline{x}(c) = \frac{1}{T}\Sigma_{t=1}^{T} x_t(c) \quad$ and

(4.4) $s(c) = [\frac{1}{T}\Sigma_{t=1}^{T}(x_t(c) - \overline{x}(c))^2]^{1/2}.$

On the other hand, let $\overline{x}_0$ and s_0 represent the averaged return
and risk of a given reference variable or an index.   Then an $\alpha - \beta$
portfolio is a portfolio $c = (c_1, \cdots, c_p)'$ belonging to the set

(4.5) $C_{\alpha\beta} = \{ c \in C |\ \overline{x}(c) \geq \overline{x}_0 + \alpha, \ s(c) \leq s_0(1 - \beta)\}.$

Of course, it may be empty, i.e., $C_{\alpha\beta} = \phi$, for large α's and
small β's.   An interesting portfolio will be a portfolio with
the same risk as the given index and with larger averaged return,
that is, a portfolio solving the maximization problem

(4.6) $\max_{c \in C} \overline{x}(c)$ subject to $s(c) = s_0.$

Another interesting portfolio is similarly a portfolio solving the
minimization problem

(4.7) $\min_{c \in C} s(c)$ subject to $\overline{x}(c) = \overline{x}_0.$

A general $\alpha - \beta$ portfolio is a portfolio maximizing the objective function

$$(4.8) \quad \phi(c) = \lambda_1 \bar{x}(c) - \lambda_2 s(c) \quad (\lambda_1, \lambda_2 > 0),$$

(see Chapter 6) or

$$(4.9) \quad \phi(c) = \bar{x}(c)/s(c).$$

To solve such a nonlinear optimization problem, we need a computer software for nonlinear optimization.

It is remarked that $\bar{x}(c) \geq \bar{x}_0 + \alpha$ is equivalent to

$$(4.10) \quad \sum_{i=1}^{P} c_i(S_{iT} - a S_{i1}) > 0 \quad \text{with} \quad a = \exp(T\bar{x}_0 + \alpha T)$$

and $s(c) \leq (1-\beta) s_0 \equiv b$ is equivalent to

$$(4.11) \quad \frac{1}{T}\sum_{t=1}^{T} x_t(c)^2 \leq \bar{x}(c)^2 + b^2.$$

5 Canonical correlation portfolio

As many empirical results have indicated, an individual stock (or asset) return series $\{x_{it}\}$ ($i=1, \cdots, p$) may not be likely to be linearly explained by a set of fundamental variables $\{z_{jt}\}$ ($j=1, \cdots, q$). However, if we make a linear combination of x_{it}'s;

$$(5.1) \quad u_t = u_t(a) = \sum_{i=1}^{P} a_i x_{it}, \quad a = (a_1, \cdots, a_p)' \quad (a_i \geq 0),$$

which is regarded as one-period ahead return of a portfolio (see Chapter 6), then the regression of u_t on z_{jt}'s

$$(5.2) \quad u_t = \beta_0 + \beta_1 z_{1t} + \cdots + \beta_q z_{qt} + w_t$$

tends to have a higher determination (or correlation) coefficient. In this section, we consider a method to construct a portfolio which maximizes the determination coefficient of regression (5.2) for given $z_{1t}, \cdots, z_{qt}$. We call it a canonical correlation port-

folio because it is an application of the canonical correlation
analysis in multivariate analysis, which is briefly reviewed later.
In the canonical correlation theory it is usually assumed that the
two groups of variables

$$(5.3) \quad x_t = (x_{1t}, \cdots, x_{pt})' \quad \text{and} \quad z_t = (z_{1t}, \cdots, z_{qt})'$$

have cross-sectional correlations but that they are serially iid.
However, the stationarity of $\{(x_t, z_t)\}$ will make the analysis
valid. In practice, it may be better to take $\widetilde{x}_{it} = S_{it}/S_{iT}$
rather than $x_{it} = \log S_{it} - \log S_{it-1}$, though $\{\widetilde{x}_{it}\}$ is unlikely to
be stationary (see Chapter 6).

Canonical correlation portfolio

Now let

$$(5.4) \quad v_t = v_t(b) = \sum_{j=1}^{q} b_j z_{jt} = b' z_t, \quad b = (b_1, \cdots, b_q)',$$

be a linear combination of fundamental variables. Then to make a
portfolio in view of the canonical correlation we maximize the correlation between u_t in (5.1) and v_t in (5.4)

$$(5.5) \quad \text{Correl}(u_t, v_t) = \frac{a' \Sigma_{xz} b}{[a' \Sigma_{xx} a \cdot b' \Sigma_{zz} b]^{1/2}}$$

under restrictions $a_i \geq 0$, where $\Sigma_{xx} = \text{Var}(x_t)$, $\Sigma_{zz} = \text{Var}(z_t)$
and $\Sigma_{xz} = \text{Cov}(x_t, z_t)$ are respectively the variance matrices of
x_t and z_t and the covariance matrix of x_t and z_t. The estimate
of (5.4) is

$$(5.6) \quad r_{uv} = r_{uv}(a, b) = \frac{a' V_{xz} b}{[a' V_{xx} a \cdot b' V_{zz} b]^{1/2}}$$

where $V_{xx} = \frac{1}{T} \sum_{t=1}^{T} (x_t - \overline{x})(x_t - \overline{x})'$, $V_{zz} = \frac{1}{T} \sum_{t=1}^{T} (z_t - \overline{z})$
$(z_t - \overline{z})'$, and $V_{xz} = \frac{1}{T} \sum_{t=1}^{T} (x_t - \overline{x})(z_t - \overline{z})'$. Thus the problem
is to maximize (5.6) with respect to a and b under $a_i \geq 0$. This
problem has always a unique solution, but because of the restric-
tion $a_i \geq 0$ we need a numerical maximization method such as Newton

Raphson method or Fletcher-Powell method. If there exists no constraint on a and b, the maximum is attained by (a_1, b_1) which satisfies

(5.7) $(V_{xz} V_{zz}^{-1} V_{zx} - \lambda_1^2 V_{xx}) a_1 = 0, \quad a_1' V_{xx} a_1 = 1,$
       $(V_{zx} V_{xx}^{-1} V_{xz} - \lambda_1^2 V_{zz}) b_1 = 0, \quad b_1' V_{yy} b_1 = 1,$

where λ_1^2 is the maximum root of $V_{xx}^{-1} V_{xz} V_{zz}^{-1} V_{zx}$. Then the maximum is

(5.8) $\max_{a, b} r_{uv}(a, b) = r_{uv}(a_1, b_1) = \lambda_1.$

If (a_0, b_0) thus obtained happens to satisfy $a_i \geq 0$, then it gives the required result.

Review on canonical correlation

Canonical correlation is an extension of the concept of correlation coefficient and measures a degree of linear relationship between two groups of random variables, say x_t and z_t as in (5.3), which are iid with mean 0 and covariances $\mathrm{Var}(x_t) = \Sigma_{11}$, $\mathrm{Var}(z_t) = \Sigma_{22}$ and $\mathrm{Cov}(x_t, z_t) = \Sigma_{12}$. To measure a correlation between x_{it}'s and z_{jt}'s as groups, we make linear combinations of x_{it}'s and z_{jt}'s;

(5.9) $u_{1t} = a_{11} x_{1t} + \cdots + a_{1p} x_{pt}, \quad v_{1t} = b_{11} z_{1t} + \cdots + b_{1q} z_{qt}$

and regard u_{1t} and v_{1t} as representatives of the x-group and z-group respectively. Then the correlation coefficient of the representatives u_{1t} and v_{1t} is given by

(5.10) $\mathrm{Correl}(u_{1t}, v_{1t}) = a_1' \Sigma_{12} b_1 / [a_1' \Sigma_{11} a_1 \cdot b_1' \Sigma_{22} b_1]^{1/2}.$

where $a_i = (a_{i1}, \cdots, a_{ip})'$ and $b_j = (b_{j1}, \cdots, b_{jq})'$ ($i, j = 1, \cdots, \min(p, q)$). This correlation coefficient depends on a_1 and b_1. In order for u_{1t} and v_{1t} to well represent each group for measuring a correlation, it is natural to choose a_1 and b_1 which jointly maximize the correlation $\mathrm{Correl}(u_{1t}, v_{1t})$. This maximization problem is equivalent to the problem:

(5.11)
$$\max_{a_1, b_1} a_1' \Sigma_{12} b_1 \quad \text{subject to} \quad a_1' \Sigma_{11} a_1 = b_1' \Sigma_{22} b_1 = 1,$$

and the solution for a_1 and b_1 is the latent vectors a_1 and b_1 of $A \equiv \Sigma_{11}^{-1} \Sigma_{12} \Sigma_{22}^{-1} \Sigma_{21}$ and $B \equiv \Sigma_{22}^{-1} \Sigma_{21} \Sigma_{11}^{-1} \Sigma_{12}$ corresponding to the largest common latent root λ_1^2. That is, let $\lambda_1^2 \geqq \lambda_2^2 \geqq \cdots \geqq \lambda_r^2$ with $r = \min(p, q)$ be the common ordered latent roots of A and B. Then a_1 and b_1 are the solutions of the following equations with $\lambda^2 = \lambda_1^2$;

(5.12) $(\Sigma_{11}^{-1} \Sigma_{12} \Sigma_{22}^{-1} \Sigma_{21} - \lambda^2 I) a = 0, \quad a' \Sigma_{11} a = 1,$
$\quad\quad\quad (\Sigma_{22}^{-1} \Sigma_{21} \Sigma_{11}^{-1} \Sigma_{12} - \lambda^2 I) b = 0, \quad b' \Sigma_{22} b = 1.$

Since the maximum value of $\text{Correl}(u_{1t}, v_{1t})$ is then $\lambda_1 (\geqq 0)$, it is called the first canonical correlation (coefficient) of the two groups, and u_{1t} and v_{1t} with a_1 and b_1 thus determined are called the first canonical variates of the x-group and z-group respectively.

Next, to measure a remaining correlation between the two groups which is not captured by the first representatives u_{1t} and v_{1t}, we make the second representatives

(5.13) $u_{2t} = a_2' x_t = a_{21} x_{1t} + \cdots + a_{2p} x_{pt}$ and
$\quad\quad\quad v_{2t} = b_2' z_t = b_{21} z_{1t} + \cdots + b_{2q} z_{qt}$

and measure the correlation

(5.14) $\text{Correl}(u_{2t}, v_{2t}) = a_2' \Sigma_{12} b_2 / [a_2' \Sigma_{11} a_2 \cdot b_2' \Sigma_{22} b_2]^{1/2}.$

However, this correlation of the second representatives should not measure the correlation captured by the first representatives through $\text{Correl}(u_{1t}, v_{1t}) = \lambda_1$. In other words, u_{2t} and v_{2t} should be "independent" representatives, carrying independent information, and they should have no correlation with u_{1t} and v_{1t}.

Hence we impose the following conditions on u_{2t} and v_{2t};

(5.15) $\mathrm{Correl}(u_{2t}, u_{1t}) = 0, \quad \mathrm{Correl}(u_{2t}, v_{1t}) = 0,$
        $\mathrm{Correl}(v_{2t}, u_{1t}) = 0, \quad \mathrm{Correl}(v_{2t}, v_{1t}) = 0.$

Thus the second representatives u_{2t} and v_{2t} are determined as those jointly maximizing $\mathrm{Correl}(u_{2t}, v_{2t})$ under the conditions (5.15), which are given by u_{2t} and v_{2t} with a_2 and b_2 satisfying the equations in (5.12) for $\lambda^2 = \lambda_2^2$. In fact, λ_2 is the maximum of $\mathrm{Correl}(u_{2t}, v_{2t})$ under (5.15) and it is called the second canonical correlation (coefficient). Also u_{2t} and v_{2t} thus obtained are called the second canonical variates of the two groups.

The third canonical variates $u_{3t} = a_3' x_t$ and $v_{3t} = b_3' z_t$ are similarly obtained as those maximizing $\mathrm{Correl}(u_{3t}, v_{3t})$ under the condition that each of u_{3t} and v_{3t} is uncorrelated with any of u_{1t}, u_{2t}, v_{1t} and v_{2t}. Again the solutions for a_3 and b_3 are given as the solutions of the equations in (5.12) with $\lambda^2 = \lambda_3^2$.

This procedure ends at $r = \min(p, q)$ because all the linear correlations on each group can be represented by the variates of the same dimension.

The above argument is based on the population parameter Σ. Since Σ is unknown in general, we replace it by $V = (V_{ij})$ ($i, j = x, z$) and all the arguments made above go through with Σ replaced by V. When (x_t, z_t)'s are iid or when $\{(x_t, z_t)\}$ is stationary, V is reasonable as an estimate of Σ.

Exercises

1. Construct an MTV index (tracking) portfolio for the reference variable you choose.

2. Also construct an index portfolio by the cluster analysis approach.

3. Construct an index $\alpha - \beta$ portfolio and compare it with those of 1 and 2.

1   Introduction

Basic concepts

There are two main types of options traded in market: European and American. Recently "Asian" options are traded over the counter (see Chapter 13). In each type, there are two classes of options; call and put. A *European call option* is a security which gives the bearer the right to buy a specified security S (underlying security) at a specified price K (exercise price or strike price) on a specified date T (maturity). An *American call option* differs from the European call only in the point that the right can be exercised anytime before or on the maturity date.

A *European put option* is a security which gives the bearer the right to sell a specified security S (underlying security) at a specified price K (exercise price or strike price) on a specified date T (maturity). An *American put option* differs from the European put option in the point that the right can be exercised anytime before or on the maturity date.

As notation we use

t : present time at which options are traded,

S_t: the price of an underlying security (e.g., stock, bond, exchange rate, etc.) at time t,

K : exercise (or strike) price specified in an option,

T : maturity date specified in an option, where the option is issued at $t=0$,

$\tau = T - t$: the remaining time for an option. Of course, the value of the option becomes null after τ.

We will treat European options.

(1) European call

A European call option can be viewed as a security (contingent claim) which pays at time T the bearer the amount

(1.1) $X_T = \max(S_T - K, 0)$ $(\geqq 0)$.

In fact, if the price of the underlying security is greater than the exercise price, i.e., $S_T > K$, then the bearer of the call option can buy the underlying security at price K by exercising the right and sell it immediately in the market to get the gain $S_T - K > 0$. If $S_T \leqq K$, he will lose $S_T - K \leqq 0$ when he exercises the right, and hence he will let the option (contract) expire. In practice, there are the following problems;

1) trading cost such as broker's fee, tax, etc. will be charged when the bearer sells the security in the market, and
2) he or she may fail to sell it at S_T because of a time lag or no trading.

These problems can be resolved when the options are listed in a market and designed in such a way that the writer (issuer) of a call option is obliged to pay the bearer the amount $S_T - K$ when $S_T > K$.

A European call is thus viewed as a security to pay the amount $X_T = \max(S_T - K, 0)$ at time T. Here since S_T is the unrealized price of an asset S at future time T, X_T is uncertain and is regarded as a random variable as S_T is a random variable. Option pricing theory aims to price a security with uncertain face value X_T. If the face value X_T is known and the interest rate r is regarded as a constant for the remaining period $\tau = T - t$, then the present value of the security is evaluated as

(1.2) $X_T^* = \exp(-r\tau) X_T.$

But at present time t X_T is unrealized (random variable), and hence X_T^* in (1.2) can not be evaluated at t.

(2) European put

Similarly a European put option is viewed as a security with maturity face value

(1.3) $X_T = \max(K - S_T, 0)$ $(\geqq 0)$.

A fundamental problem in option theory is to price a contingent claim X_T theoretically. Let C_t and P_t be the prices (premiums) of the contingent claims X_T in (1.1) and X_T in (1.3) respectively. C_t and P_t depend on variables S_t, K and τ at least and they will also depend on interest rate r. We write, for example, $C_t = C_t(K)$ when C_t is a function of K with remaining variables fixed.

Fundamental pricing lemma via arbitrage

Option pricing theory is based on the concept of arbitration, not on the concept of equilibrium on which economics is based. In economics, equilibrium prices are considered to be prices which equate demands and supplies, which are explained through behaviors of economic agents. Hence, specification of a concept of economic agents, specification of utility functions, specification of expectations (forecasts) formation of economic agents, etc. will be needed. On the other hand, arbitrage theory does not need such specifications, and it simply uses the basic concept that profitable opportunities without risk eventually disappear. Relationships between arbitrage theory and equilibrium theory are not well studied, but a situation with an arbitrage opportunity is not an equilibrium. Here an arbitrage opportunity is a profitable opportunity with no risk and no initial fund. In capital asset markets, asset prices are immediately mutually adjusted in order that no arbitrage opportunity is available.

Some basic conditions for such arbitrage theory to hold will be (i) market trading, (ii) free entry to market, and (iii) perfect information on asset prices.

Now let A_t and B_t be the prices of assets a and b at time t, which may be composite assets (or financial products).

Assumption

i) Assets a and b are traded in market and any long or short posi-

tion on a and b can be taken without limitation where any size
of cash loan is available with a fixed interest rate r (com-
pleteness of market),

ⅱ) No trading cost is charged, i.e., no broker's fee, no tax , no
margin requirement, etc. (frictionless market),

ⅲ) Market is perfect.

Fundamental Lemma (1) If the prices of a and b at future time T
($> t$) satisfy

(1.4) $A_T \geqq B_T$ (with probability 1),

then the prices at each t satisfies

(1.5) $A_t \geqq B_t$,

(2) If $A_T = B_T$ at T ($> t$) with probability 1, then it holds at
 t that

(1.6) $A_t = B_t$.

Proof.   (1) If $A_t < B_t$, then short b at price B_t and long a at
price A_t and reverse the positions at T .   Then the following pay-
off table holds at each time.
In other words, there are a
positive profit $B_t - A_t > 0$
at t and a nonnegative profit $A_T - B_T \geqq 0$ with no
risk at T because (1.4)
holds with probability 1.

	t	T
a	$-A_t$	A_T
b	B_t	$-B_T$
total	$B_t - A_t > 0$	$A_T - B_T \geqq 0$

By assumption, such an
opportunity must disappear so that (1.5) holds.   (2) Apply
the argument in (1) to $A_T \geqq B_T$ and to $A_T \leqq B_T$ to get (1.6).

In an application of this lemma, we only need to specify assets a and b. An important result from this simple lemma is

Put-Call Parity

$$(1.7) \quad S_t + P_t = C_t + K e^{-rt}.$$

Proof.   Let $A_t = S_t + P_t$ and $B_t = C_t + K e^{-rt}$. Then it is easy to see $A_T = B_T$ via (1.1) and (1.3), because K and T in $P_t = P_t$ (K, T) and $C_t = C_t(K, T)$ are common.

The following results are also obtained directly from the lemma.

Proposition 2.1.   [Boundaries of European calls]
(a) $S_t \geq C_t \geq \max[0, S_t - K e^{-rt}]$.
(b) $C_t(K_1) \geq C_t(K_2)$ for $K_2 > K_1$.
(c) $K_2 - K_1 \geq C_t(K_1) - C_t(K_2)$ for $K_2 > K_1$.
(d)   For any α $(0 \leq \alpha \leq 1)$ and $K_2 > K_1$,
$$C_t(\alpha K_1 + (1-\alpha) K_2) \leq \alpha C_t(K_1) + (1-\alpha) C_t(K_2).$$
(e) $C_t(\tau_2) \geq C_t(\tau_1)$ for $\tau_2 > \tau_1$.

Corresponding boundaries for puts are obtained by using (1.7).

2   Black-Scholes option pricing formula

In general, an exact option theory consists of the following steps.

(I) The stochastic process of underlying asset prices on which options are written is specified;

$$S = \{ S_t : 0 \leq t \leq T \}.$$

(II) A contingent claim (option) X_T is specified as in (1.1) or (1.3).

(III) For bond B and asset S, a trading rule (stochastic process) giving portfolio units at each time t

$$\boldsymbol{b} = \{ \boldsymbol{b}_t = (b_{0t}, b_{1t}) : 0 \leq t \leq T \}$$

is specified in such a way that

(i) for the value process of its portfolio

(2.1) $V_t = V_t(\boldsymbol{b}) = b_{0t} B_t + b_{1t} S_t,$

$V_T(\boldsymbol{b}) = X_T$ (with probability one) at T and

(ii) neither inflow of money nor outflow of money are allowed in each trading (self-finance condition) (see below).

(IV) Then by Fundamental Lemma the value of the contingent claim X_T at t is equal to the value $V_t(\boldsymbol{b})$ of the portfolio at t.

In case of a call, $X_T = C_T = \max(S_T - K, 0)$ in the step (I) and in case of a put, $X_T = P_T = \max(K - S_T, 0)$. It is not quite necessary for X_T to be of these forms nor X_T needs to be a function of S_T alone. Also in (II), b_{it}'s ($i = 0, 1$) are in general functions of past and present prices $\{ S_s : 0 \leq s \leq t \}$, but not functions of future prices.

The mathematical structure of the option theory is thus written as $(\mathbb{S}, X_T, \boldsymbol{b})$.

BS Formula

In the BS option theory the above steps are specified as follows.

I   Price process $\mathbb{S} = \{ S_t : 0 \leq t \leq T \}$. The price process of an underlying asset is assumed to be a geometric Brownian motion

(2.2) $S_t = S_0 \exp\{ \sigma W_t + (r - \frac{1}{2} \sigma^2) t \},$

and the bond price process is

$B_t = \exp\{ - r(T - t) \} = \exp(- r\tau),$

where the process $\mathbb{W} = \{ W_t : 0 \leq t \leq T \}$ of W is a Wiener process, i.e., Brownian process with drift (mean) 0 and variance t (see Appendix for a Brownian process and some stochastic calculus).

II Contingent claim X_T. A European call is specified as

(2.3) $X_T = \max(S_T - K, 0)$.

III Trading rule $b = \{b_t : 0 \leq t \leq T\}$. It is necessary to specify
a self-financing trading rule b for which $V_T(b) = X_T$. It is
noted that b_{it}'s can take negative units.

IV As a consequence the BS pricing formula for a call is given by

(2.4) $C_t = S_t \Phi(d_t) - K e^{-rt} \Phi(d_t - \sigma\sqrt{\tau})$,

where

$$d_t = \frac{\log(S_t/K) + (r + \frac{1}{2}\sigma^2)\tau}{\sigma\sqrt{\tau}},$$

$$\Phi(x) = \int_{-\infty}^{x} \frac{1}{\sqrt{2\pi}} \exp(-\frac{1}{2} z^2)\, d z.$$

The practical validity of this theoretical pricing depends great
ly on the practical validity of the assumption that $\{S_t\}$ follows a
geometric Brownian motion. Since the BS formula was established,
various theoretical extensions of the formula have been made to
adapt it to practical situations. An example is that $\{S_t\}$ is assumed to be a geometric Brownian motion with Poisson jumps. But
these extensions are still theoretical adaptations and they are not
necessarily consistent with empirical features of financial return
series, such as nonnormality, non-independence, nonlinearity, etc.
In fact, the assumption of Brownian motion of $\{S_t\}$ implies that
its return series $\{x_t\}$ where $x_t = \log S_t - \log S_{t-1}$ follows an iid
normal process or equivalently $\{\log S_t\}$ follows a normal strong
random walk, which is not the case in reality, as has been observed
in Chapter 2.

On the derivation of the BS formula (2.4).

Black and Scholes themselves did not derive it through the steps I~IV, but derived it through forming a portfolio of bond, stock and call with no risk for an infinitesimal interval $[t, t+\Delta t]$. There the formula was shown to be the unique solution of the partial differential equation

(2.4) $\dfrac{1}{2}\sigma^2 S_t^2 \dfrac{\partial^2 C_t}{\partial S_t^2} + \dfrac{\partial C_t}{\partial t} + r S_t \dfrac{\partial C_t}{\partial S_t} - r C_t = 0$

with boundary condition $C_T = X_T$ in (2.3). On the other hand, in the above derivation procedure of steps I~IV, portfolio units b_{it}'s are continuously adjusted in order that the value $V_T(\boldsymbol{b})$ of portfolio (2.1) at time T be equal to the value X_T of a call at time T. An important concept in this approach is the concept of self-finance that in adjusting the portfolio ($\boldsymbol{b}_{t-\Delta t} \to \boldsymbol{b}_t$) from $t - \Delta t$ to t at t, the value of the portfolio before adjustment must be equal to the one after adjustment;

(2.6) $b_{0\,t-\Delta t} B_t + b_{1\,t-\Delta t} S_t = b_{0t} B_t + b_{1t} S_t.$

In (2.6), the left hand side is the value of a portfolio at t made at $t - \Delta t$ and held for the period $[t - \Delta t, \ t]$, while the right hand side is the value after changing $b_{it-\Delta t}$'s to b_{it}'s at t. This approach also gives the partial differential equation (2.5) (see Jarrow and Rudd (1983)). It is easy to check by substitution that the BS formula is a solution of (2.5). Also it is easy to see that $C_t \to X_T$ as $t \to T$ (Problem 8).

Risk neutrality and its implications

The BS formula in (2.4) is shown to be equal to the present value of the expected value of the contingent claim (call);

(2.7) $C_t = e^{-r\tau} E_t(X_T) = e^{-r\tau} E[\max(S_T - K, 0) | \mathcal{S}_t],$

where expectation is conditional on $\mathcal{S}_t - \{S_s : 0 \leqq s \leqq t\}$. This

result is often associated with the risk neutrality argument that
if people ignore the risk (variation) factor of a contingent claim
X_T, then they will evaluate it by the discounted value of its mean
$E_t(X_T)$ (risk neutrality evaluation). In reality, we are not in
the world of risk-neutrality and hence we cannot ignore the risk
factor of X_T. Therefore even if this argument is an interpretation of the BS formula, it will not be the pricing theory for a
contingent claim X_T. However the observation in (2.7) contains
useful contents for the following two problems.

1) In the above I~Ⅳ, a certain self-financing trading rule gives
$V_t(\boldsymbol{b}) = C_t$ and $V_T(\boldsymbol{b}) = X_T$. Hence (2.7) is expressed as

$$(2.8) \quad E[\, e^{-rT} V_T(\boldsymbol{b}) \,|\, \mathscr{S}_t] \;=\; e^{-rt} V_t(\boldsymbol{b}),$$

implying that the discounted value process defined by

$$\{\, e^{-rt} V_t(\boldsymbol{b}) : 0 \leqq t \leqq T\}$$

is a martingale (MG) process with respect to $\{\mathscr{S}_t\}$. This suggests
that the MG property of the value process of a self-financing portfolio will be deeply related to the possibility of pricing a contin
gent claim. This viewpoint is fully implemented by Harrison and
Pliska (1981). They developed an option theory in a more general
framework with martingale (MG) stochastic integral and for example,
obtained the following result.

Theorem 2.1. (Harrison and Pliska (1981)) When $\mathscr{S} = \{\, S_t : 0 \leqq t \leqq T\}$ follows a geometric Brownian motion (2.1), any contingent claim
$X_T = \phi(S_T)$ which is a function of S_T is uniquely priced by the
present value $e^{-rt} E_t(X_T)$ of the conditional expected value of
X_T given $\mathscr{S}_t$.

It is noted that this result is not based on the risk neutrality
argument but on the replication of X_T by the procedure I~Ⅲ

described above. According to this theorem, such a nonlinear contingent claim as $X_T = (S_T - K)^2$ can be theoretically priced as $e^{-r\tau} E_t(X_T)$.

2) A price process S in reality such as stock price process, exchange rate process, etc. will not follow a geometric Brownian motion, as has been observed in Chapter 2 and hence we may evaluate the price of an option X_T by the discounted expected value $e^{-r\tau} E_t$ (X_T) as a proxy. This is a compromized solution for the theoretical limitation and a practical necessity. This standpoint will be taken in Chapter 13 and options are evaluated under price processes described by Taylor model, ARCH model, etc.

Rates of change of option premiums

The BS (theoretical) call and put option prices (premiums) depend on the five variables;

(2.9) $S_t, \ K, \ \sigma, \ \tau = T - t$ and r.

The theory assumes the constancy of volatility σ and interest rate r of a "safe" asset, while exercise price K is constant by its definition of option. However, except for exercise price K, σ and r change in reality. In fact, as has been observed in Chapter 2, the volatilities represented by absolute returns $|\, x_t - \bar{x} \,|$ or $|\, x_t \,|$ fluctuate greatly and are serially correlated, which implies that the price process $\{S_t\}$ is not a geometric Brownian motion on which the BS theory is developed. On the other hand, underlying asset price S_t and time to maturity τ certainly change at each moment. Hence it is natural to consider the sensitivities of option prices to these variables. Let us first consider the sensitivities of a call to the five variables. Let $C_t(S_t, \ \sigma, \ \tau, \ K, \ r)$ be the BS call premium. Then a change of call premium due to changes of these variables is given by

$$(2.10) \quad \Delta C_t = \frac{\partial C_t}{\partial S_t} \Delta S_t + \frac{1}{2} \frac{\partial^2 C_t}{\partial S_t^2} (\Delta S_t)^2 + \frac{\partial C_t}{\partial \sigma} \Delta \sigma$$
$$+ \frac{\partial C_t}{\partial \tau} \Delta \tau + \frac{\partial C_t}{\partial r} \Delta r + \frac{\partial C_t}{\partial K} \Delta K$$

which is a Taylor expansion of the BS call premium with respect to the five variables in (2.9). In (2.10), the second partial derivative $\partial^2 C_t / \partial S_t^2$ is included because S_t changes greatly and C_t is a highly nonlinear function of S_t. Since K cannot change in reality, set $\Delta K = 0$ in applications. The partial derivatives of the BS call and put premiums are termed and computed as follows.

(1) Delta (Δ)

$$\Delta = \frac{\partial C_t}{\partial S_t} = \Phi(d_t), \qquad \Delta = \frac{\partial P_t}{\partial S_t} = -\Phi(-d_t).$$

(2) Gamma (Γ)

$$\Gamma = \frac{\partial^2 C_t}{\partial S_t^2} = \frac{\Phi'(d_t)}{S_t \sigma \sqrt{\tau}}, \qquad \Gamma = \frac{\partial^2 P_t}{\partial S_t^2} = \frac{\Phi'(d_t)}{S_t \sigma \sqrt{\tau}}.$$

(3) Kappa (κ)

$$\kappa = \frac{\partial C_t}{\partial \sigma} = S_t \sqrt{\tau} \, \Phi'(d_t), \quad \kappa = \frac{\partial P_t}{\partial \sigma} = S_t \sqrt{\tau} \, \Phi'(d_t).$$

(4) Theta (Θ)

$$\Theta = \frac{\partial C_t}{\partial \tau} = \frac{S_t \sigma}{2\sqrt{\tau}} \Phi'(d_t) + r K \Phi(d_t - \sigma \sqrt{\tau}) \, e^{-r\tau},$$
$$\Theta = \frac{\partial P_t}{\partial \tau} = \frac{S_t \sigma}{2\sqrt{\tau}} \Phi'(d_t) - r K \Phi(-d_t + \sigma \sqrt{\tau}) \, e^{-r\tau}.$$

(5) Roh (ρ)

$$\rho = \frac{\partial C_t}{\partial r} = \tau K e^{-r\tau} \Phi(d_t - \sigma \sqrt{\tau}),$$
$$\rho = \frac{\partial P_t}{\partial r} = -\tau K e^{-r\tau} \Phi(-d_t + \sigma \sqrt{\tau}).$$

Using this notation, ΔC_t in (2.10) with $\Delta K = 0$ is expressed as

$$\Delta C_t = \Delta(\Delta S_t) + \Gamma(\Delta S_t)^2 + \kappa \Delta \sigma + + \Theta \Delta \tau + \rho \Delta r.$$

3 Basic formulae for BS option portfolios

For notation, t denotes the present time up to which asset prices $\{S_s: s \le t\}$ are observed, and for a random variable Z_T to be realized at T its discounted (present) value is denoted by

$$(3.1) \quad Z_T^* = e^{-r\tau} Z_T \quad \text{where} \quad \tau = T - t.$$

Under this notation $C_t = E_t[C_T^*]$ and $S_t = E_t[S_T^*]$. In this section we derive some basic formulae for option portfolios in the framework of the BS world. An important point is that from a theoretical viewpoint C_t is only the expected value at t of a random variable $X_T^* = C_T^*$ as a function of S_T.

Theorem 3.1. Assume the geometric Brownian motion (1.1) for $\{S_t\}$ with $\mu = r - \frac{1}{2}\sigma^2$.

(1) $\text{Var}(C_T^*) = S_t \bar{C}_t - K^* C_t - C_t^2$
$$= S_t[\bar{C}_t - C_t] - C_t P_t \quad \text{with}$$

$$\bar{C}_t = S_t \exp(\sigma^2 \tau) \Phi(d_\tau + \sigma \sqrt{\tau}) - K^* \Phi(d_\tau).$$

(2) $\text{Var}(P_T^*) = S_t(\bar{P}_t + P_t) - C_t P_t \quad \text{with}$

$$\bar{P}_t = S_t \exp(\sigma^2 \tau)[1 - \Phi(d_\tau + \sigma \sqrt{\tau})] - K^*[1 - \Phi(d_\tau)].$$

(3) $\text{Var}(S_T^*) = S_t^2[\exp(\sigma^2 \tau) - 1].$

(4) $\text{Cov}(S_T^*, C_T^*) = S_t(\bar{C}_t - C_t).$

Proof. Outlined in Appendix.

Corollary 3.1. (1) The correlation coefficient $\rho_{sc} = \text{Correl}(S_T^*, C_T^*)$ of S_T^* and C_T^* is given by

$$\rho_{sc} = \frac{\bar{C}_t - C_t}{\{[\exp(\sigma^2 \tau) - 1][S_t(\bar{C}_t - C_t) - C_t P_t]\}^{1/2}}.$$

(2) Let $\sigma_c^2 = \text{Var}(C_T^*)$ and $\sigma_s^2 = \text{Var}(S_T^*)$. The portfolio of the form $a S_t + (1 - a) C_t$ at time t which minimizes $\Lambda(a) = \text{Var}(a S_T^*$

$+(1-a)\,C_T^*)$ is given by

$$(3.3) \qquad a_0 = \frac{\sigma_s\sigma_c\rho_{sc} - \sigma_c^2}{\sigma_s^2 - 2\sigma_s\sigma_c\rho_{sc} + \sigma_c^2}$$

and its minimum is given by

$$(3.4) \qquad \Lambda(a_0) = \frac{\sigma_c^2[4\sigma_c^2 - 8\sigma_s\sigma_c\rho_{sc} + \sigma_s^2(3\rho_{sc}^2 + 1)]}{\sigma_s^2 - 2\sigma_s\sigma_c\rho_{sc} + \sigma_c^2} .$$

The variances of call and put options are important because
a) options are objects for investments and their risks can be
   measured by their variances, and
b) BS option pricing formula simply gives an approximate value in
   practice because prices do not follow geometirc Brownian motions,
   and hence it will be useful to have a measure of deviation from
   the theoretical value even though the measure is derived under
   the same assumption.
The distribution of C_T^* may be a more direct measure of risk because it gives probabilities that C_T^* deviates from the mean $C_t = E_t(C_T^*)$.

Corollary 3.2. When $\{S_t\}$ is a geometric Brownian motion, the
distribution of C_T^* is given by

$$(3.5)$$
$$F(z) = P(C_T^* \le z) = \begin{cases} \Phi\left(\frac{1}{\sigma\sqrt{\tau}}\log\left[\frac{K + e^{r\tau}\,z}{S_t}\right] - \frac{\mu\sqrt{\tau}}{\sigma}\right) & (z \ge 0), \\ 0 & (z < 0), \end{cases}$$

where $\mu = r - \frac{\sigma^2}{2}$. In particular,

$$(3.6) \qquad F(0) = P(C_T^* = 0) = \Phi\left(-\frac{1}{\sigma\sqrt{\tau}}\log(S_t/K) - \frac{\mu\sqrt{\tau}}{\sigma}\right) > 0.$$

Proof is outlined in Appendix.

    $F(0)$ in (3.5) is the probability that a call security becomes a
trash, and hence the smaller, the better. $F(0)$ is a decreasing
function of S_t/K, and an increasing function of σ.   If σ is

given, $F'(0)$, $\mathrm{Var}(C_T^*)$ in Theorem 3.1 (1) and $P(C_T^* > C_t)$ are evaluated.

Recall that the derivative of a call option with respect to volatility σ, which is called option kappa κ, is given by

(3.7) $\kappa = \partial C_t / \partial \sigma = \partial E_t[C_T^*] / \partial \sigma = S_t \sqrt{\tau} \, \Phi'(d_t).$

This is the rate of change of the expected (mean) value of C_T^*, with respect to σ. Similarly it is natural to consider the rate of change of the variance of C_T^* with respect to σ;

(3.8) $\dfrac{\partial \mathrm{Var}(C_T^*)}{\partial \sigma} \equiv 2\, S_t^2 \sigma \, \tau \exp(\sigma^2 \tau)\, \Phi(d_t + \sigma\sqrt{\tau}) - 2\, K C_t.$

We call this option kappa of the variance, while we call (3.7) option kappa of the mean. The measure (3.8) can be important in option portfolio context.

Vertical option portfolios

It is known that many option positions can be constructed by simply combining different options. A position of options with different strike prices but with the same maturity is called a vertical position. A general vertical portfolio including the underlying asset at time T is expressed as

(3.9) $X_T^* = \sum_{i=1}^{m} [\alpha_i C_T^*(K_i) + \beta_i P_T^*(K_i) + \gamma_i S_T^*]$

with $K_1 > K_2 > \cdots > K_m$, where $C_T^*(K_i)$ and $P_T^*(K_i)$ are respectively discounted call and put with strike price K_i. The conditional mean X_T^* given $\{S_s : s \leq t\}$ is clearly

(3.10) $E_t[X_T^*] = \sum_{i=1}^{m} [(\alpha_i + \beta_i) C_t(K_i) + (\gamma_i - \beta_i) S_t + \beta_i K_i],$

where we used the put-call parity $P_T^*(K_i) + S_T^* = C_T^*(K_i) + K_i^*$. On the other hand, the variance of X_T^* is evaluated as

(3.11)

$$\mathrm{Var}(X_T^*) = \sum_{j=1}^{m} \sum_{i=1}^{m} (\alpha_i + \beta_i)(\alpha_j + \beta_j)[S_t \overline{C}_{t,i} - K_i^* C_{t,j} - C_{t,i} C_{t,j}]$$
$$+ \lambda^2 S_t^2[\exp(\sigma^2 \tau) - 1] + 2\lambda \sum_{i=1}^{m} (\alpha_i + \beta_i) S_t[\overline{C}_{t,i} - C_{t,i}],$$

where $C_{t,i} = C_t(K_i),$

$\overline{C}_{t,i} = S_t \exp(\sigma^2 \tau) \, \Phi(d_{i,t} + \sigma \sqrt{\tau}) - K_i^* \Phi(d_{i,t}),$

$d_{i,t} = [\log S_t / K_i + (r + \frac{1}{2}\sigma^2)\tau] / \sigma \sqrt{\tau}$ and

$\lambda = \sum_{i=1}^{m} (\gamma_i - \beta_i)$

(see Appendix for an outline of proof). Note that

(3.12) $\mathrm{Cov}(C_T^*(K_i), C_T^*(K_j)) = S_t \overline{C}_{t,i} - K_i^* C_{t,j} - C_{t,i} C_{t,j}$ ($i < j$).

Example 1. Long strangle and long straddle

When $m = 2$, $\alpha_1 = 0$, $\beta_1 = 1$, $\gamma_1 = 0$, $\alpha_2 = 1$, $\beta_2 = 0$, and $\gamma_2 = 0$, the position X_T^* in (3.9) is called a strangle;

$$X_T^* = P_T^*(K_1) + C_T^*(K_2)   (K_1 > K_2)$$

Its mean is $E(X_T^*) = P_t(K_1) + C_t(K_2)$ and the variance of X_T^* is given by

$$\mathrm{Var}(X_T^*) = \sum_{j=1}^{2} \sum_{i=1}^{2} [S_t \overline{C}_{t,i} - K_i^* C_{t,j} - C_{t,i} C_{t,j}]$$
$$+ S_t^2[\exp(\sigma^2 \tau) - 1] - 2 \sum_{i=1}^{2} S_t[\overline{C}_{t,i} - C_{t,i}].$$

When $m = 1$, $\alpha_1 = 1$, $\beta_1 = 1$ and $\gamma_1 = 0$, the position is called a straddle in which a large deviation from strike price K_1 will yield a profit, whether S_t may rise or fall so long as it moves away from K_1. The position is expressed as $X_T^* = C_T^*(K_1) + P_1^*(K_1)$. Hence its mean is $E(X_T^*) = C_t(K_1) + P_t(K_1)$ and the variance of X_T^* is given by

$$\mathrm{Var}(X_T^*) = 4[S_t \overline{C}_{t,1} - K_1^* C_{t,1} - C_{t,1}^2] + S_t^2[\exp(\sigma^2 \tau) - 1]$$
$$- 2 S_t[\overline{C}_{t,1} - C_{t,1}].$$

4 Portfolio Insurance

In general, a portfolio insurance is a portfolio strategy or a trading rule which makes the value of an asset or a portfolio at t more than or equal to a prespecified level after a certain period. The strategy is often realized by forming a portfolio combined with options or option equivalents. A typical example is either (1) or (2) in the followings.

(1) One unit Stock + one unit Put with strike price K.

(2) One unit Call with strike price K + K unit bond with maturity at T.

These are simply the left and right sides of the put call parity

$$(4.1) \quad S_t + P_t = C_t + K e^{-rt}$$

respectively and the value of the left portfolio (and hence the right portfolio) is not less than K at T. In fact

$$(4.2) \quad S_T + P_T = S_T + \max(K - S_T, 0) \geq K.$$

Hence holding a put with strike price K in addition to an underlying security S keeps the value of S not less than the lower bound K (floor) at time T. The cost of forming this insured portfolio is the premium $P_t(K)$ of the put with strike price K, which will be totally charged when $S_T \leq K$. If $S_T > K$ at time T, the cost is reduced to $P_t - (S_T - K)$, and if it is negative, the capital gain completely compensates the payment of $P_t(K)$. In this section, we consider a fundamental PI (portfolio insurance) problem in the framework of the BS option theory.

PI problem

At time t, a fund W_t is invested on γ units of a stock with price S_t and γ units of a put for S with premium $P_t(K)$, exercise price K and maturity T so that W_t is invested on the insured portfolio

(4.3) $W_t = \gamma [S_t + P_t(K)] \equiv \gamma D_t(K)$.

Then at time T the asset value (wealth) becomes

(4.4) $W_T = \gamma [S_T + P_T(K)] = \gamma [\max(S_T, K)] \geq \gamma K$,

and hence the portfolio (4.3) has lower bound γK at T. But since put price $P_t(K)$ at t is positive, the lower bound γK will be smaller than the initial fund W_t. In other words, as K is set to be larger, γ will become smaller so that γK will remain less than W_t, though we wish to have a perfectly insured portfolio with $W_T \geq W_t$ (with probability 1). Our problem is thus to maximize the lower bound

(4.5) $F(K) = \gamma K = K W_t / D_t(K)$

with respect to strike price K subject to budget constraint (4.3) where W_t is fixed. Note that γ in (4.3) and (4.5) is a function of K. Of course, our world is assumed to be the world of the BS option theory where $\{S_t\}$ follows a geometric Brownian motion and an European put is priced by

(4.6) $P_t(K) = -S_t \Phi(-d_t) + K e^{-r\tau} \Phi(-d_t + \sigma\sqrt{\tau})$ with

(4.7) $d_t \equiv d_t(S_t/K) = \left[\log(S_t/K) + (r + \frac{1}{2}\sigma^2)\tau\right]/\sigma\sqrt{\tau}$.

Theorem 4.1.   There exists no strike price K which maximizes the lower bound $F(K)$ in (4.5). In fact, $F(K)$ is an increasing continuous function of K satisfying

(4.8) $F(K) = \gamma K < e^{r\tau} W_t$ and $\lim_{K \to \infty} F(K) = e^{r\tau} W_t$.

Proof.   Differentiating (4.5) with respect to K yields

$$F'(K) = W_t[S_t - S_t \Phi(-d_t)]/D_t(K)^2 > 0.$$

where the L'Hopital's theorem was used in the first equality.

Some implications of this result are as follows.

i) The lower bound (floor) of the insured portfolio (protective put) at time T is less than the value $e^{rt} W_t$ which is obtained from the investment of the fund W_t on a bond with interest rate r and maturity T.

ii) For any K less than $e^{rt} W_t$, there exists (K, γ) such that $W_t = \gamma [S_t + P_t(K)]$ and $W_T \geq \gamma K$. But since $P_t(K)$ is an increasing function of K, if γK is set to be close to $e^{rt} W_t$, or equivalently if K is large (relative to S_t), the put premium $P_t(K)$ becomes close to $K e^{-rt}$ so that

(4.9) $\gamma \equiv \gamma(K) = W_t / [S_t + P_t(K)]$

becomes very small.   Therefore if we demand a stronger insurance by setting a higher floor γK, the capital gain over the floor γK at time T, i.e.,

(4.10) $W_T - \gamma K = \gamma [S_T + \max(K - S_T, 0)]$
$$= \gamma(K)[S_T - K] = \gamma(K) S_T - F(K)$$

becomes smaller.   In fact, this capital gain over γK is a strictly decreasing function of K.   This implies that there is a trade-off between the capital gain $W_T - \gamma K$ and the insured floor γK. Of course, put options with large K's are not available in practice.

We study two interesting cases; (a) $\gamma K = W_t$ and (b) $K = S_t$ (at-the-money strike price).

Case (a); $\gamma K = W_t$.   By Theorem 4.1, such (K, γ) uniquely exists.

Theorem 4.2.   Let $x = S_t / K$.   The (K, γ) for which $\gamma K = W_t$ is given by $(\widetilde{K}, \widetilde{\gamma})$ with $\widetilde{K} = S_t / \widetilde{x}$ and $\widetilde{\gamma} = W_t / \widetilde{K} = \widetilde{x} W_t / S_t$ where $\widetilde{x}$ is the unique solution of the equation

(4.11) $1 - x + x \, \Phi(- d_\tau(x)) - e^{-r\tau} \Phi(- d_\tau(x) + \sigma\sqrt{\tau}) = 0.$

Furthermore $\tilde{x} = S_t / \tilde{K} < e^{-r\tau}$ and in particular $S_t < \tilde{K}.$

Proof. Substituting $\gamma = W_t / K$ into (4.3) yields (4.11). Denoting the left side of (4.11) by $f(x)$, we obtain

$$f'(x) = -1 + \Phi(- d_\tau(x)) < 0.$$

Hence $f(x)$ is strictly decreasing in x and it follows that

$$\lim_{x \to 0} f(x) = 1 - e^{-r\tau} \quad \text{and} \quad \lim_{x \to \infty} f(x) = -\infty.$$

This implies that $f(x) = 0$ has a unique solution $\tilde{x} > 0$. Further setting $x = e^{-r\tau}$ and using $\Phi(-\frac{1}{2}\sigma\sqrt{\tau}) = 1 - \Phi(\frac{1}{2}\sigma\sqrt{\tau})$, $f(x) = 1 - 2\Phi(\frac{1}{2}\sigma\sqrt{\tau}) < 0$, implying $\tilde{x} < e^{-r\tau}$. In particular $S_t < \tilde{K}$, completing the proof.

An alternative form of (4.11) is

(4.12) $1 - e^{-r\tau} = x \, \Phi(d_\tau(x)) - e^{-r\tau} \Phi(d_\tau(x) - \sigma\sqrt{\tau}).$

The right side of (4.12) is the BS call premium with strike price 1 (dollar), while the left side is the difference between the maturi- ty value (1 dollar) of the bond which pays 1 dollar at T and its present value $e^{-r\tau}$. Hence we shall call $\tilde{x} = x(\sigma, \tau)$ satisfying (4.11) or equivalently (4.12) a bond-call parity value for $x = S_t / K.$

Case (b); $K = S_t$.   In this case, by (a) the lower bound γK is less than the initial fund W_t, because $\gamma K = W_t$ implies $K > S_t$ and because γK is increasing in K. When $K = S_t$, $\gamma K = W_t / g$ by (4.3) where

$$g = g(\sigma, \tau) = (1 + e^{-r\tau}) \Phi(\tfrac{1}{2}\sigma\sqrt{\tau}) + 1 - e^{-r\tau}.$$

Hence γK is independent of $K = S_t$. For a given τ, $g(\sigma, \tau)$ is increasing in volatility σ and hence γK is decreasing in σ. If σ is very large, g is close to 2 so that lower bound γK is close to $W_t/2$. In other words, the value of the insured portfolio (4.4) with at-the-money put can end up at maturity with the half of the initial value W_t in the worst case, which can happen when $\frac{1}{2}\sigma\sqrt{\tau}$ is close to 2.5, that is, σ is close to $5/\sqrt{\tau}$. If $\frac{1}{2}\sigma\sqrt{\tau} \doteqdot 1$, or equivalently $\sigma = 2/\sqrt{\tau}$, $g \doteqdot 1.65 - 0.35\,e^{-r\tau} > 1.3$ and so $\gamma K = W_t/g < W_t/1.3 \doteqdot 0.77$, implying that the initial fund can be reduced to the level of 77%. Of ourse, γK is close to W_t when σ is close to zero, but in that case there will be no strong incentive to forming the insured portfolio.

5 Estimation of volatilities

In the BS theory, it is assumed that price process $\{S_t\}$ obeys a geometric Brownian motion with volatility (standard deviation) σ constant. On the other hand, it is empirically confirmed that volatilities in many return series are likely to change. Hence to estimate the volatility parameter σ in the BS formula, it is often practiced that

(1) σ is estimated based on recent data (small sample) as

(5.1)　　$s = [\Sigma_{s=t-m}^{t}(x_s - \bar{x})^2/m]^{1/2}$　with　$x_t = \log S_t - \log S_{t-1}$ or

(2) a volatility series as defined as $\{|x_t - \bar{x}|\}$ is modelled and
　　predicted by a time series model as in Chapter 3, or

(3) a series of implied volatilities derived from the market option
　　premiums is modelled and predicted as in (2).

These practices are not well consistent with the basic structure of the BS theory, but they are often undertaken. While, theories with volatilities changing have been developed in a line with the BS theory. However, these theoretical formulations for volatility processes are rather ad hoc and do not have empirically solid bases.

Patell and Wolfson (1979) empirically observed that
1) options with larger volatilities tend to be overpriced (over-
   valued) and options with smaller volatilities tend to be underpriced, and
2) in stock options implied volatilities of options with longer
   life are more variable than those of options with shorter life.
In the literature, assuming the framework of the BS theory, many
efficient estimation procedures which may use high, low, opening
prices in addition to closing prices have been proposed. Here we
consider the case where closing prices are available.

Implied volatilities at the money

Let $\widetilde{C}_t$ be the at-the-money market price at t of a call option
with τ time to maturity and let $C_t(\sigma)$ denote the BS theoretical
call price as a function of σ. Since $C_t(\sigma)$ is a monotone increasing function of σ, for each time the equation

(5.2) $\widetilde{C}_s = C_s(\sigma)$ $(s=t, t\text{-}1, \cdots, t\text{-}m)$

yields a unique value $\widetilde{\sigma}_s = C_s^{-1}(\widetilde{C}_s)$. Hence one may model the
series $\{\widetilde{\sigma}_s\}$ or $\{\log\widetilde{\sigma}_s\}$ of implied volatilities by such a time
series model as an AR(p) model, and predict the volatility at matu-
rity $t + \tau = T$. However, if the prices $\widetilde{C}_s$ are overpriced or
underpriced for any reasons, the predicted value will be biased
even if the BS formula is practically effective. Whaley (1982) and
Latane and Rendleman (1976) proposed different estimation procedures for implied volatilities.

Statistical modelling for $\{|x_t - \bar{x}|\}$

When we regard $|x_t - \bar{x}|$ as an estimate of volatility at t, we
can model it by such a time series model as described in Chapter 3.
An example for Nikkei 225 is given there. As is observed, the auto-
correlations of $\{|x_t - \bar{x}|\}$ are in general slowly declining and

positive.   Hence process $\{|\,x_t-\bar{x}\,|\}$ may be of such long-term dependency as will be discussed in Section 4 of Chapter 13.   In that case, one may use a fractional ARIMA(p,d,q) model for the process, and predicted values from this model may be used for the BS formula.   More directly one may evaluate the expected value of a contingent claim via Monte Carlo method under a FARIMA model (see Section 4 of Chapter 13).

EXercises

1. Using Fundamental Lemma, prove Proposition 2.1.

2. Using a formula in Appendix, show that (2.1) is equivalent to the usual expression in finance textbooks

$$d\,S_t/\,S_t\;=\;r\,d\,t+\sigma\,d\,W_t.$$

3. Show that the BS call premium in (2.4) is by substitution a solution of the partial differential equation (2.5) with (2.3).

4. Show that when $\log S_T \sim \mathrm{N}(\log S_t+\mu\,\tau,\,\sigma^2\tau)$ with $\mu=r-\sigma^2/2$, $e^{-r\tau}\,E(X_T)$ is C_t in (2.3) where $X_T=\max(S_T-K,0)$.

5. Verify the partial derivatives Δ, Γ, κ, Θ and ρ in Section 2.

6. With help of Appendix, give the details of the proof of Theorem 3.1 and Corollaries 3.1 and 3.2.

7. Give the details of the proof of Theorem 4.1.

8. Show that C_t in (2.4) converges to X_T in (2.2) as $t\to T$.

CHAPTER 13

PRACTICAL OPTION PRICING AND RELATED TOPICS

1 Practical option pricing

The BS (Black-Scholes) option theory and its generalized version given by Harrison and Pliska (1981) provide mathematically beautiful and powerful results on option pricing. However, as has been observed in Chapters 2 and 3, a price process $\{S_t\}$ in reality will not satisfy the assumptions the theory requires. In fact, the time series structure of return series $\{x_t\}$ with $x_t = \log S_t - \log S_{t-1}$ will not admit a (dominating) measure with respect to which the discounted process $\{e^{-rt}S_t\}$ becomes a martingale. Hence we will not be able to develop an arbitrage pricing theory by forming an equivalent (replicated) portfolio. In other words, in such a non-linear model as Taylor model which is consistent with empirical features of returns, it is difficult to replicate an option and price it by the fundamental lemma in Chapter 12. As has been discussed, in such a case we are interested in the distribution of the present value $X_T{}^* = \exp(-r\tau)X_T$ of a contingent claim X_T and often regard the expected value $E(X_T{}^*)$ as a proxy for pricing maybe with help of a risk neutrality argument, where $E(\cdot)$ is evaluated at t and $\tau = T - t$. In this case, evaluations of the variance $\mathrm{Var}(X_T{}^*)$ and of such probabilities as $P(X_T{}^* = 0)$, $P(X_T{}^* \geq E(X_T{}^*))$ etc. will be also important for practical purposes, as discussed in Chapter 12. In this chapter we consider such practical pricing problems and related topics in some specific models.

Basic formulation

A practical pricing of a European call needs
(1) a specification of return process $\{x_t\}$ or
(2) a specification of the distribution of S_T.

242

In fact, the value of a European call option at maturity T is given by

(1.1) $X_T = \max(S_T - K, 0) = \max(S_t \exp(z_\tau) - K, 0)$,

where $\tau = T - t$ and

(1.2) $z_\tau = x_{t+1} + \cdots + x_{t+\tau}$ with $x_t = \log S_t - \log S_{t-1}$.

Here, whether or not price process $\{S_t\}$ may be a martingale, we practically price X_T by its discounted expected value;

(1.3) $C = E(X_T^*) = \exp(-r\tau) E_t[X_T]$,

in which we may also evaluate the variance $\text{Var}(X_T^*)$ to obtain a risk measure for C as we discussed in Chapter 12.   If we can identify or specify the time series process of $\{x_t\}$, we can in principle evaluate the mean of X_T.   The evaluation may be based on
   a) an analytical approach or
   b) a Monte Carlo simulation approach.
In analytical approach the expected value of X_T^* is attempted to be mathematically evaluated with respect to the specified process. In Section 3, assuming that $\{x_t\}$ follows a Taylor model, this procedure will be taken.   However, even in such an analytical approach, to get a value of C it is necessary to evaluate it in a numerical method because $E(X_T^*)$ is usually a rather complicated function just like the BS formula.   On the other hand, in Monte Carlo approach, time paths for $\{x_{t+1}, \cdots, x_{t+\tau}\}$ are repeatedly generated with respect to the model for process $\{x_t\}$ by computer and for each path the sum z_τ in (1.2) is computed to generate the distribution of $S_T = S_t \exp(z_\tau)$ and hence of X_T.   In other words, generating thousands of time paths, we can obtain an approximate distribution of $X_T^* = e^{-r\tau} X_T$, the mean and variance of X_T^*, estimates of $P(X_T^* = 0)$, $P(X_T^* > a)$ etc.   If necessary, the skewness and kurtosis of the distribution are obtained.   This Monte

Carlo approach is rather powerful not only for the case that an analytical approach is too complicated but also for the case that various exotic options which may be time-dependent have to be priced.

Practical impossibility of option pricing under nonlinear models

As has been discussed in Chapter 3, a nonlinear model with conditional heteroscedasticity can be developed by specifying conditional normal distribution for return at t:

$$(1.4) \quad x_t \quad \text{given} \quad \psi_{t-1} \sim N(h_t, g_t^2),$$

where $h_t = h(\psi_{t-1})$, $g_t^2 = g(\psi_{t-1})^2$ and $\psi_{t-1} = \{S_s | s \leq t-1\}$. For example, in the GARCH model the conditional mean h_t and conditional variance g_t^2 in (1.4) are respectively specified as

$$(1.5) \quad h_t = \phi_0 + \phi_1 x_{t-1} + \cdots + \phi_p x_{t-p} \quad \text{and}$$
$$g_t^2 = \alpha_0 + \Sigma_{i=1}^q \alpha_i \varepsilon_{t-i}^2 + \Sigma_{j=1}^r \beta_j g_{t-j}^2$$

where $\varepsilon_t = x_t - h_t$. Threshold autoregressive (TAR) model, smoothing transition model, etc. treated in Chapter 3 are also special cases of the model (1.4).

Now recall that a condition for a European call to be priced theoretically is by Theorem in Chapter 12 that $\{e^{-rt}S_t\}$ forms a martingale process (with respect to a measure which is absolutely continuous with respect to Lebesgue measure). Here r is a fixed interest rate.

Lemma 1.1   A necessary and sufficient condition for $\{e^{-rt}S_t\}$ to be a martingale process in the model (1.4) is that $h_t = r - \frac{1}{2}g_t^2$.

Proof.   When x_t follows (1.4),

$$E[e^{-rt}S_t | \psi_{t-1}] = E[e^{-rt}S_{t-1}\exp(x_t) | \psi_{t-1}]$$
$$= e^{-rt}S_{t-1}\exp(h_t + \frac{1}{2}g_t^2),$$

which is equal to S_{t-1} if and only if $h_t = r - g_t^2/2$.

Therefore unless the conditional mean h_t is equal to $r - \frac{1}{2}g_t^2$, the discounted price process $\{e^{-rt}S_t\}$ cannot be a martingale (with respect to the normal measure (1.4)). This implies that the GARCH (1.4) with (1.5) but without $h_t = r - \frac{1}{2}g_t^2$ cannot be a martingale unless $\alpha_i = 0$ ($i = 1, \cdots, p$) and $\beta_j = 0$ ($j = 1, \cdots, q$), i.e., unless $\{S_t\}$ is a geometric Brownian motion. However, a model given by

$$(1.6) \quad x_t \quad \text{given} \quad \psi_{t-1} \sim N\left(r - \frac{1}{2}g_t^2, \; g_t^2\right)$$

is always a martingale whatever $g_t = g(\psi_{t-1})$ may be. Thus a GARCH-like model (1.6) with g_t^2 in (1.5) is a martingale. In such a model a European type option X_T is priced by the present value of its expected value $e^{-rt} E(X_T)$.

In reality, it is very unlikely that the conditional mean $h_t = h(\psi_{t-1})$ of x_t is equal to $r - \frac{1}{2}g_t^2$ with $g_t^2 = g(\psi_{t-1})^2$ being equal to the conditional variance of x_t. If $h_t = r - \frac{1}{2}g_t^2$. larger conditional variations would tend to yield negative returns even if prices are greatly rising. In other words, the nonlinear time series structure in (1.1) will not get along with the martingale property.

2 Gram-Charlier Option pricing
The iid case

From the argument in Section 1, we are interested in deriving the distribution of $X_T^* = \exp(-r\tau)X_T$. But it is obviously difficult to specify the distribution of $S_T = S_t \exp(z_\tau)$ without knowledge of the process of $\{x_t\}$ where $z_\tau = \sum_{h=1}^{\tau} x_{t+h}$ as in (1.2). If we could assume that x_t's are iid (independently and identically distributed) with mean $E(x_t) = \mu$ and variance $Var(x_t) = \sigma^2$, then $E(z_\tau) = \tau\mu$ and $Var(z_\tau) = \tau\sigma^2$ and $\sqrt{\tau}(\bar{z} - \mu)$ may be approximated by normal distribution $N(0, \sigma^2)$ for τ large, where $\bar{z} = z_\tau/\tau$. That is, if x_t's are iid and if

τ is large, z_τ may be approximated by $N(\tau \mu, \tau \sigma^2)$, in which case the call premium C in (1.3) is approximately equal to the BS value C^{BS}. However, as has been observed in Chapter 2, even if x_t's are iid, the skewness β and kurtosis κ of $\{x_t\}$ are not empirically equal to those of normal distribution. The information will be better to be taken into the pricing of a contingent claim X_T. To do so, it is easy to see that the skewness and kurtosis of z_τ under $\{x_t\}$ iid are respectively given by

$$(2.1) \quad \beta_\tau = \beta/\sqrt{\tau} \quad \text{and} \quad \kappa_\tau = 3(1-\tau^{-1})+\kappa/\tau.$$

When τ is large, $\beta_\tau \sim 0$ and $\kappa_\tau \sim 3$. But when τ is not large, at least κ_τ may not be regarded as 3 because sometimes κ is empirically very large (see the case of Nikkei 225 in Chapter 2). While, β is usually small. Therefore it will be better to approximate the distribution of z_τ by a distribution with mean $\mu_\tau = \tau \mu$, variance $\sigma_\tau^2 = \tau \sigma^2$, skewness β_τ and kurtosis κ_τ which is close to $N(\mu_\tau, \sigma_\tau^2)$ in some sense. Such a distribution is given by a special case of the Gram-Charlier expansion with density

$$(2.2) \quad g(z) = f(z) - \frac{1}{3!}\beta_\tau \sigma_\tau^3 f^{(3)}(z) + \frac{1}{4!}\eta_\tau \sigma_\tau^4 f^{(4)}(z),$$

where $\eta_\tau = \kappa_\tau - 3$, $f(z)$ is the density of $N(\mu_\tau, \sigma_\tau^2)$ and $f^{(i)}(z)$ is the i-th derivative of $f(z)$. In fact, $g(z)$ satisfies the properties;

1) $g(z) \geq 0$ and $\int g(z)\,dz = 1$,
2) $E(z) = \mu_\tau = \tau \mu$ and $\text{Var}(z) = \sigma_\tau^2 = \tau \sigma^2$,
3) the skewness and kurtosis of $g(z)$ are β_τ and κ_τ respectively.

We evaluate the mean of the present value of a European call option

$$(2.3) \quad X_T = \max(S_t \exp(z_\tau) - K, 0),$$

based on the density in (2.2). Let $X_T^* = e^{-r\tau} X_T$ as before and let

(2.4) $H_i(d_1) = \int_{-\infty}^{d_1} (z - \sqrt{\sigma_\tau})^i (2\pi)^{-1/2} \exp(-z^2/2) \, dz,$

(2.5) $L_i(d_2) = \int_{-\infty}^{d_2} z^i (2\pi)^{-1/2} \exp(-z^2/2) \, dz,$

(2.6) $d_1 = [\log S_t/K + \mu_\tau + \sigma_\tau^2]/\sigma_\tau$ and $d_2 = d_1 - \sigma_\tau.$

Lemma 2.1   Let $a_1 = \exp(-r\tau)$ and $a_2 = \exp(\mu_\tau + \frac{1}{2}\sigma_\tau^2).$ Then

(2.7) $E[X_T^*] = G_0 - \frac{1}{3!}\beta_\tau G_3 + \frac{1}{4!}\eta_\tau G_4,$

where $\eta_\tau = \kappa_\tau - 3$, $L_0(d_2) = H_0(d_2),$

$$G_0 = a_1[a_2 S_t H_0(d_1) - K L_0(d_2)]$$

(2.8) $G_3 = a_1[a_2 S_t H_3(d_1) - K L_3(d_2)]$
$\qquad\qquad - 3 a_1[a_2 S_t H_1(d_1) - K L_1(d_1)]$

$\qquad G_4 = a_1[a_2 S_t H_4(d_1) - K L_4(d_2)]$
$\qquad\qquad - 6 a_1[a_2 S_t H_2(d_1) - K L_2(d_1)] + 3 G_0.$

Proof.   Transform z into $w = z - \mu_\tau$ with $X_T^* = a_1 \max(a_2 S_t \exp(w) - K, 0)$. Then the density of w is $g(w)$ and $f^{(3)}(w) = [-(w/\gamma)^3 + 3(w/\gamma)] f(w)$ and $f^{(4)}(w) = [(w/\gamma)^4 + 6(w^2/\gamma^3) + (3/\gamma^2)] f(w)$ with $\gamma = \sigma_\tau^2$. Hence evaluating $E(X_T^*)$ yields the result.

If $\mu = r - \frac{1}{2}\sigma^2$, then $a_1 a_2 = 1$ so that G_0 equals the BS value;

(2.9) $G_0 = C^{BS} = S_t \Phi(d_1) - K e^{-r\tau} \Phi(d_2).$

However, it is not guaranteed in reality that $\mu = r - \frac{1}{2}\sigma^2$. We call $E(X_T^*)$ in (2.7) the Gram-Charlier option premium. Of course, if $\beta_\tau = 0$ and $\eta_\tau = \kappa_\tau - 3 = 0$, (2.7) is reduced to G_0.

In Jarrow and Rudd (1983) a similar attempt has been made. But in their approach the distribution of S_T itself is approximated by an Edgeworth series, which is unnatural.

A non iid case

Without modelling the time series structure of return process $\{x_t\}$, one may evaluate the discounted expected value $E(X_T{}^*)$ of a European call option in the same way as in the iid case. The basic assumption is that $\{x_t\}$ is a 4-th order stationary process in the sense that in addition to $E(x_t)=\mu$ and $\mathrm{Cov}(x_t, x_{t+k})=\gamma_k$ for all t,

(2.10) $\delta_{hk}=E(\widetilde{x}_t \widetilde{x}_{t+h} \widetilde{x}_{t+k})$ is independent of t
for all h and k,

(2.12) $\lambda_{hkm}=E(\widetilde{x}_t \widetilde{x}_{t+h} \widetilde{x}_{t+k} \widetilde{x}_{t+m})$ is independent of t
for all h, k and m,

where $\widetilde{x}_t=x_t-\mu$. Under this assumption the mean, variance, skewness and kurtosis of $z_\tau=\Sigma_{h=1}^\tau x_{t+k}$ are respectively

(2.13)

$$\mu_\tau = \tau\mu, \qquad \sigma_\tau^2 = \tau\gamma_0+2\Sigma_{h=1}^\tau(\tau-h)\gamma_h$$

$$\sigma_\tau^3\beta_\tau = \tau\delta_{00}+3[\Sigma_{h=1}^\tau(\tau-h)\delta_{hh}+\Sigma_{h=1}^\tau(\tau-h)\delta_{0h}]$$
$$+6[\Sigma_{k=2}^\tau(\tau-k)\delta_{1k}+\Sigma_{k=3}^\tau(\tau-k)\delta_{2k}+\cdots+\delta_{\tau-2\ \tau-1}]$$

$$\sigma_\tau^4\kappa_\tau = \tau\lambda_{000}+4[\Sigma_{h=1}^\tau(\tau-h)\lambda_{00h}+\Sigma_{h=1}^\tau(\tau-h)\lambda_{hhh}]$$
$$+12\Sigma_{k=1}^\tau(\tau-k)\lambda_{0kk}+24[\Sigma_{k=3}^\tau(\tau-k)\lambda_{12k}$$
$$+\Sigma_{k=4}^\tau(\tau-k)\lambda_{23k}+\cdots+\lambda_{\tau-3\ \tau-2\ \tau-1}].$$

These four moments of z_τ are estimated by estimating μ, $\{\gamma_k\}$, $\{\delta_{jk}\}$ and $\{\lambda_{ijk}\}$ based on past observations of x's. For example, for $i\leq j\leq k$

$$\hat{\lambda}_{ijk}=\Sigma_{t=1}^{n-k}(x_t-\overline{x})(x_{t+i}-\overline{x})(x_{t+j}-\overline{x})(x_{t+k}-\overline{x})/n.$$

Now similarly to the iid case, we may approximate the distribution of z_τ by the distribution with density $g(z)$ in (2.2). Then by

Lemma 2.1, $E(X_T^*)$ is evaluated by (2.7) with σ_τ^2, β_τ and κ_τ in
(2.12). It should be noted that μ is very unlikely to be equal to
$r - \frac{1}{2}\sigma_\tau^2$ because σ_τ^2 is given by (2.12) (see also Section 1).

Consequently if we can assume the 4-th order stationarity for
return process $\{x_\iota\}$, the distribution of z_τ may be approximated
by the Gram-Charlier density (2.2) through which an option is evaluated as its discounted expected value.

3 Practical option pricing under a Taylor model and under an ARCH model

As has discussed in Section 1, once a model is specified for
returns $\{x_\iota\}$, it is possible to evaluate the expected value (and
variance) of a contingent claim X_T by generating thousands of time
paths via the model. However, there always remains an uncertainty
about the solidity of a result. In other words, the result depends
to some extent on each Monte Carlo simulation, though it may be
rather stable if the number of repetitions is large. On the other
hand, sometimes an (approximately) analytical evaluation of $E(X_T)$
can be made when a model for returns is simple.

A. Taylor(1986)'s method

To describe the evaluation of a call proposed by Taylor (1986),
we start with

Lemma 3.1. Let $\log S_T \sim N(\gamma, \delta^2)$ and consider a call option X_T
$= \max(S_T - K, 0)$. Then

$$E(X_T) = \exp(\gamma + \frac{1}{2}\delta^2)\, \Phi\left(\frac{\gamma + \delta^2 - \log K}{\delta}\right) - K\Phi\left(\frac{\gamma - \log K}{\delta}\right).$$

A Taylor model discussed in Chapter 2 is of the form

$$(3.1) \quad x_\iota - \mu = u_\iota v_\iota \quad \text{with} \quad x_\iota = \log S_\iota - \log S_{\iota-1},$$

and hence in this model the price S_T at T is expressed as

(3.2) $\log S_T = \log S_t + \mu \tau + \Sigma_{h=1}^{\tau} u_{t+h} v_{t+h}.$

In the below we treat the following cases separately.

(3.3)   Case i) $\{u_t\}$ iid $N(0,1)$.

        Case ii) $\{u_t\}$ is of a probabilistic trend.

Also it is assumed that $\{v_t\}$ follows log AR(1) process;

(3.4) $\log v_{t+h} - \alpha = \phi\{\log v_{t+h-1} - \alpha\} + \eta_t,$
         $\eta_t \sim N(0, \beta^2(1-\phi)).$

Case i): $\{u_t\}$ iid $N(0,1)$. Then by (3.3)

(3.5) $\log S_T$ given $\{v_t\} \sim N(\gamma, \delta_0^2)$ with
         $\gamma = \log S_t + \mu \tau$ and $\delta_0^2 = \Sigma_{h=1} v_{t+h}^2.$

Taylor proposed to approximate (3.5) by

(3.6) $\log S_T \sim N(\gamma, \delta^2)$ with $\delta^2 = \Sigma_{h=1}^{\tau} E[v_{t+h}^2 | v_t].$

If this approximation is accepted, δ^2 in (3.6) is evaluated as

(3.7) $\delta^2 = \Sigma_{h=1}^{\tau} \exp\{2\alpha + 2\phi^h[\log v_t - \alpha] + 2\beta^2(1-\phi^{2h})\}$

since by (3.4) conditional on v_t

(3.8) $\log v_{t+h} \sim N(\alpha + \phi^h[\log v_t - \alpha], \beta^2(1-\phi^{2h})).$

Therefore by Lemma 3.1 and (3.6), the present value of the conditional expectation of X_T given v_t is evaluated as

(3.9) $C_t = S_t \Phi \left(\frac{1}{\delta}[\log(S_t/K) + r\tau + \frac{1}{2}\delta^2]\right)$
         $- K \Phi \left(\frac{1}{\delta}[\log(S_t/K) + r\tau - \frac{1}{2}\delta^2]\right),$

where Taylor specified $\mu = r - \frac{1}{2\tau}\delta^2$. To estimate δ, letting m_s $= |x_s - \overline{x}|$ and using degrees-of-freedom-adjusted autocorrelation

(3.10) $r_{k, m}' = n r_{k, m}/(n - k)$ with

$r_{k, m} = \sum_{s=k+1}^{n} (m_s - \overline{m})(m_{s-k} - \overline{m})/\sum_{s=1}^{n} (m_s - \overline{m})^2,$

first β and ϕ are estimated by minimizing

$$\sum_{k=1}^{K} (r_{k, m}' - \rho_{k, m}(\beta, \phi))^2,$$

where n is the sample size up to time point t (typically $n = t$, $n-1 = t-1, \cdots, 1 = t - n+1$). Here $\rho_{k, m}(\beta, \phi)$ is the k lag autocorrelation of $m_s = |x_s - \mu|$, evaluated under (3.4) as

(3.11) $\rho_{k, m} = \dfrac{[2\exp(\beta^2)-1][\exp(\beta^2 \phi^k)-1]}{[\pi\exp(\beta^2)-2][\exp(\beta^2)-1]}$.

Also α is estimated by $\hat{\alpha} = \sum_{s=1}^{n} \log \hat{v}_s / n$ with $\hat{v}_s = 1.253 m_s$. Hence an estimate of δ is obtained by substituting these $\hat{\alpha}$, $\hat{\beta}$, $\hat{\phi}$ and $\hat{v}_s$ into (3.7), and C_t in (3.9) is finally estimated. This is Taylor's procedure.

B. Generalization of Taylor's method

In the above evaluation, it is assumed that $\{u_t\}$ is iid $N(0,1)$ and (3.5) is "approximated" by (3.6). Here we treat a general case: $\{u_t\}$ is normally stationary and we formulate an evaluation formula without approximation. For this purpose, let

(3.12) $u = (u_T, u_{T-1}, \cdots, u_{t+1})': \tau \times 1,$
         $u^* = (u_t, u_{t-1}, \cdots, u_{t-k})': (k+1) \times 1,$
         $v = (v_T, v_{T-1}, \cdots, v_{t+1})': \tau \times 1,$
         $v^* = (v_t, v_{t-1}, \cdots, v_{t-k})': (k+1) \times 1.$

Since $\{u_t\}$ is normally stationary, (u, u^*) follows a $(\tau + k+1)$ dimensional multivariate normal distribution $N(0, \Omega)$. Hence for $\Omega = (\Omega_{ij})$ with $\Omega_{12}: \tau \times (k+1)$ $(i, j = 1, 2)$,

(3.13) u given $u^* \sim N(B u^*, \Sigma)$ with
         $B = \Omega_{12} \Omega_{22}^{-1}$ and $\Sigma = \Omega_{11.2} = \Omega_{11} - \Omega_{12} \Omega_{22}^{-1} \Omega_{21}.$

Therefore the conditional distribution of $\log S_T$ given u^* and (v, v^*) is

(3.14) $\log S_T$ given (u^*, v, v^*) $\sim$ $N(\gamma, \delta^2)$ with
$$\gamma = \log S_t + \mu \tau + v' B u^* \quad \text{and} \quad \delta^2 = v' \Sigma v.$$

Applying Lemma 3.1 to (3.14) yields the conditional evaluation of a call X_T;

(3.15) $H(u^*, v) \equiv e^{-r\tau} E[X_T | u^*, v, v^*] = e^{-r\tau} \times (3.1),$

where γ and δ^2 in (3.1) are replaced by those in (3.14). Consequently if u^* and v^* are observable, an evaluation of a call is given by

(3.16) $\Delta(u^*, v^*) = E[H(u^*, v) | v^*].$

To evaluate this formula, it is necessary to sepcify the process of $\{v_t\}$. Taylor proposed $\log$ AR(1) in (3.4). Even if we take it, we need to evaluate (3.16) numerically by computer. As a process of $\{u_t\}$, one may use such a simple process as AR(1), MA(1), etc. We leave this problem to the readers.

Another way to evaluate (3.16) is as follows.

(1) Specify a model for $\{u_t\}$ and a model for $\{v_t\}$ by the method of Chapter 3.
(2) Using past observations, estimate the models.
(3) Generate thousands of time paths for $\{u_t\}$ and $\{v_t\}$ through the models to estimate B and $\delta^2 = \delta^2(v)$ for each v and to evaluate the expected value in (3.16).

Practical option pricing under an ARCH model

In a simple ARCH model $x_t = \rho x_{t-1} + \varepsilon_t$ with

(3.17) ε_t given $\psi_{t-1} \sim N(0, \alpha_0 + \alpha_1 \varepsilon_{t-1}^2)$

(see Chapter 3 for ARCH model), let us consider the problem of evaluating the expected value of a European call $X_T = \max(S_t \exp (z_t) - K, 0)$ where $z_t = \sum_{h=1}^{\tau} x_{t+h}$. First note

(3.18) $x_t = \varepsilon_{t+h} + \rho \varepsilon_{t+h-1} + \cdots + \rho^{h-1} \varepsilon_{t+1} + \rho^h x_t$

and hence

(3.19)
$$z_t = \sum_{h=1}^{\tau} x_{t+h} = \varepsilon_{t+\tau} + (1+\rho) \varepsilon_{t+\tau-1} + (1+\rho+\rho^2) \varepsilon_{t+\tau-2}$$
$$+ \cdots + (1+\rho+\cdots+\rho^{\tau-1}) \varepsilon_{t+1} + \rho (1+\rho+\cdots+\rho^{\tau-1}) x_t.$$

In the model (3.17) it is very difficult to evaluate $E(X_T)$ analytically. Here, supposing that ρ, α_0 and α_1 are estimated, we consider a Monte Carlo approach. For given x_t and x_{t-1}, $\varepsilon_t^2 = (x_t - \rho x_{t-1})^2$ is obtained and hence an ε_{t+1} is generated from $N(0, \alpha_0 + \alpha_1 \varepsilon_t^2)$. Next we generate ε_{t+2} from $N(0, \alpha_0 + \alpha_1 \varepsilon_{t+1}^2)$ for the generated ε_{t+1}. Repeating the procedure, a time path $\{\varepsilon_{t+1}, \varepsilon_{t+2}, \cdots, \varepsilon_{t+\tau}\}$ is generated. Therefore, we are able to generate thousands of time paths of $\{\varepsilon_t\}$, through which z_t's in (3.19) are generated. Hence $E(X_T)$ can be evaluated as

(3.20) $\overline{X}_T = \dfrac{1}{N} \sum_{j=1}^{N} X_{Tj} = \dfrac{1}{N} \sum_{j=1}^{N} \max(S_t \exp(z_{tj}) - K, 0)$

where $X_{Tj} = \max(S_t \exp(z_{tj}) - K, 0)$ and z_{tj} is the value of z_t for the j-th generation of time paths of $\{\varepsilon_t\}$. It should be noted that when $\alpha_1 \varepsilon_t^2$ happens to be large, there are some possibilities that $\varepsilon_{t+\tau}^2$ is very large. In other words, an estimate of $\mathrm{Var}(X_T)$ given by

(3.21) $\dfrac{1}{N} \sum_{j=1}^{N} (X_{Tj} - \overline{X}_T)^2$

may be very large. This is a property of an ARCH model (see Chapter 3).

4 Fractional Brownian motion and option pricing

In Section 2, it is observed that though return process $\{x_t\}$ itself does not exhibit a strong autocorrelation structure, its absolute return process $\{|x_t|\}$ carries a strong, slowly declining and positive autocorrelation structure. This implies that $\{x_t\}$ is not iid and hence that price process $\{S_t\}$ is not a geometric Brownian motion. Recently such a slowly declining autocorrelation of $\{|x_t|\}$ is sometimes associated with the long-term dependency or equivalently long memory of a stationary time series. In general, a stationary process with autocovariances γ_j's is said to be of a long-term dependency if

$$(4.1) \quad \Delta \equiv \Sigma_{j=0}^{\infty} |\gamma_j| = \infty.$$

This condition of course implies that γ_j's for j large do not diminish quickly like a stationary AR model. If $\Delta < \infty$, $\{z_t\}$ has a continuous spectral density $f(\lambda)$ and $f(\lambda)$ at $\lambda = 0$ converges absolutely to

$$(4.2) \quad f(0) = \gamma_0 + 2\Sigma_{j=0}^{\infty} \gamma_j.$$

In other words, if $\Delta = \infty$, $f(0)$ does not converge absolutely and hence the long term dependency of a stationary process as defined by (4.1) is related to the variations of low frequencies (long cycles and trend movements). It is remarked that even if $\Delta = \infty$, $\{z_t\}$ may have a spectral density. To treat such a long-term dependency,

a) Fractional ARIMA(p, d, q) model or simply FARIMA(p, d, q) model

b) Fractional Brownian motion or simply FBG(H) model

have been proposed by Granger and Joyeux (1980) (also Hosking (1981)) and Mandelbrot and Van Ness (1968) respectively. In the sequal we briefly describe these models and introduce an application to option pricing due to Takahashi (1992).

a) Fractional ARIMA(p, d, q) model

A FARIMA(p, d, q) model is defined to be an ARIMA(p, d, q) model where d can be any real number, contrary to the usual ARIMA model where d is a nonnegative integer. That is, $\{z_t\}$ follows a FARIMA (p, d, q) model if $y_t = (1-L)^d z_t$ follows an ARMA(p, q) model where $L^k z_t = z_{t-k}$ and

$$(4.3) \qquad (1-L)^d = \Sigma_{k=0}^{\infty} \binom{d}{k}(-L)^k \quad \text{with} \quad d \in R.$$

In the usual ARIMA(p, d, q) model analysis, such a difference as $y_t = (1-L)z_t = z_t - z_{t-1}$ is taken to secure the stationarity of the differenced series $\{y_t\}$ because $\{z_t\}$ may be nonstationary. However, sometimes such a procedure yields an over-defferenced series where its spectral density gets negligible weights at low frequencies and variations of high frequencies are more weighted. To overcome such an overdifference problem, Granger and Joyeux (1980) and Hosking (1981) proposed a stationary FARIMA(p, d, q) model with $|d| < \frac{1}{2}$. In fact, differently from the usual ARIMA case, $z_t = (1-L)^{-d} y_t$ itself forms a stationary process if $|d| < \frac{1}{2}$ where $\{y_t\}$ follows an ARMA(p, q) model, and its autocovariances are of order:

$$(4.4) \qquad \gamma_k \sim |k|^{2d-1} \quad \text{as} \quad k \to \infty \quad (\lim_{k \to \infty} \gamma_k / |k|^{2d-1} = \text{const}).$$

Hence if $0 < d < \frac{1}{2}$, $\Delta = \Sigma_{k=0}^{\infty} |\gamma_k| = \infty$ and so $\{z_t\}$ is of a long-term dependency. If $-\frac{1}{2} < d < 0$, γ_k diminishes quickly and it is shown that the spectral density $f(\lambda)$ gets 0 at $\lambda = 0$ so that the model may be suitable for a phenomenon of big variations at high frequencies. It is noted that a stationary FARIMA(p, d, q) model also carries a short-term dependency inherited from the ARMA structure in

$$(4.5) \qquad z_t = \Sigma_{k=0}^{\infty} \binom{-d}{k}(-1)^k y_{t-k} \quad \text{with} \quad \{y_t\} \text{ ARMA}(p, q).$$

In particular, if $\{y_t\}$ is a white noise $\{\varepsilon_t\}$ and $0 < d < \frac{1}{2}$ or equivalently if $\{z_t\}$ follows a FARIMA$(0, d, 0)$ with $0 < d < \frac{1}{2}$, z_t is a weighted average of the white noise as

$$(4.6) \quad z_t = \sum_{k=0}^{\infty} w_k \varepsilon_{t-k} \quad \text{with} \quad w_k = \binom{-d}{k}(-1)^k = \frac{\Gamma(k+d)}{\Gamma(d)\,k!}.$$

This process with normality for $\{\varepsilon_t\}$ is sometimes regarded as a model for a return process $\{x_t\}$. However, since the autocovariances of $\{x_t\}$ are usually small and die out very qucikly, it will not be a good model for $\{x_t\}$ itself. Rather, it may be a suitable model for absolute return process $\{|x_t|\}$.

Fractional Brownian motion FBM(H)

A Gaussian (normal) stochastic process $\{B_t : 0 \leq t\}$ which satisfies

$$E(B_t) = 0 \quad \text{and} \quad E(B_t - B_s)^2 = c(t-s)^{2H} \quad (t \geq s)$$

is called a fractional Brownian motion with parameter H where H is the initial for "Hurst". It is shown that

(1) if $H = \frac{1}{2}$, $\{B_t\}$ is a Brownian motion,

(2) the distribution of $B_t - B_0$ is equal to that of $t^H(B_1 - B_0)$, which is called a self-similar property,

(3) $\text{Var}(B_t) = t^{2H} \text{Var}(B_1)$, and

(4) when $\frac{1}{2} < H < 1$, $z_t = B_t - B_{t-1}$ forms a stationary process whose spectral density of the form

$$(4.7) \quad f(\lambda) = f^*(\lambda)/|1 - \exp(i\lambda)|^{2d} \quad \text{where} \quad d = H - \frac{1}{2}$$

where $f^*(\lambda) > 0$ for $\lambda \in (-\pi, \pi)$ (Geweke, Porter and Hudak (1983)).

The process $\{z_t\}$ in (4) is said to be a fractional Gaussian noise (FGN). From (4), $\{z_t\}$ has autocorrelations satisfying

(4.8) $\gamma_k \sim |k|^{2d-1}$ as $k \to \infty$ with $d = H - \frac{1}{2}$,

hence $\{z_t\}$ is a process of a long-term dependency. This implies that the autocorrelations for k large behave like those of a stationary FARIMA(p, d, q) model whose spectral density is of the form

(4.9) $f(\lambda) = f_0(\lambda)/|1 - \exp(i\lambda)|^{2d}$,

where $f_0(\lambda)$ is the spectral density of the ARMA(p, q) model. In other words, the long term behaviors between a FGN and a FARIMA model are not well distinguished. Mandelbrot and Van Ness (1968) proposed the Fractional Brownian motion associated with the concept of self-similarity. Sometimes log-price process $\{\log S_t\}$ is assumed to follow a fractional Brownian motion, in which returns x_t's follow a FGN(H) model.

Application to option pricing

In Takahashi (1992), it is assumed that a return process $\{x_t\}$ follows a FGN(H) model and an approximate premium for an European call option is evaluated through an argument of the Edgeworth expansion as

(4.12) $C_t = S_t \Phi(d_1) - K\exp(-r\tau)[\Phi(d_2) - \delta_2]$

with

(4.13) $d_1 = [\log(S_t/K) + r\tau + \frac{1}{2}\sigma^2\tau^{2H} + \log\delta_1]/\sigma\tau^H$

$d_2 = [\log(S_t/K) + r\tau + \frac{1}{2}\sigma^2\tau^{2H} + \log\delta_1]/\sigma\tau^H$,

$\delta_1 = 1 + \frac{1}{2}e_1(\sigma\tau^H)^2 - \frac{1}{6}e_3(\sigma\tau^H)^3 + \frac{1}{24}e_4(\sigma\tau^H)^4$, and

$\delta_2 = \frac{1}{2}e_2(\sigma\tau^H)\phi(d_2) + \frac{1}{6}e_3(\sigma\tau^H)\phi^{(1)}(d_2) - (\sigma\tau^H)^2\phi(d_2)]$

$\quad + \frac{1}{24}e_4[(\sigma\tau^H)\phi^{(2)}(d_2) - (\sigma\tau^H)^2 g^{(1)}(d_2) + (\sigma\tau^H)^3\phi(d_2)]$,

where $\Phi(z) = \int_{-\infty}^z \phi(x)\,dx$, $\phi(z) = (2\pi)^{-1/2}\exp(-z^2/2)$, $\phi^{(i)}(z) = d^i\phi(z)\,dz^i$ and

(4.14) $e_i = \kappa_i^* - \kappa_i$ with $\kappa_i = \int_{-\infty}^{\infty} z^i \phi(z) \, dz$

and κ_i^*'s are moments under FGN(H) model. The idea for derivation
of (4.12) is similar to Jarrow-Rudd (1983) where the distribution
of S_T itself (not x_T) is expanded around a normal distribution by
Edgeworth expansion. Takahashi (1992) observed through a numerical
analysis that for $H > \frac{1}{2}$ the call premiums under FGN model at the
money are smaller than those of the BS call. In applications it is
required to estimate σ^2 and the Hurst parameter (exponent) H.

Estimation of H — R/S analysis

Hurst (1951) found out through so-called R/S analysis that $x_t = B_t - B_{t-1}$ is not iid normal in the analysis of long term storage
capacity of water reserviors. He used the rescaled range statistic;

(4.15) $Q_n = R_n / s_n = [\max_k \Sigma_{j=1}^{k} \widetilde{x}_j - \min_k \Sigma_{j=1}^{k} \widetilde{x}_j] / s_n$

where $\widetilde{x}_j = x_j - \overline{x}_n$, $\overline{x}_n = \Sigma_{j=1}^{n} x_j / n$ and $s_n^2 = \Sigma_{j=1}^{n} \widetilde{x}_j^2 / n$. If
$\{x_t\}$ is iid normal, it is shown that $Q_n/\sqrt{n}$ converges in distri-
bution to the range V of Brownian bridge on $[0,1]$ defined by

(4.16) $V = \max_t BB_t - \min_t BB_t$

where $BB_t = W_t - t W_1$ with $\{W_t\}$ a BM(0,1) on $[0,1]$. It is also
shown that

(4.17) $E(V) = (\pi/2)^{1/2}$ and $Var(V) = \pi^2/6$.

In the usual R/S analysis proposed by Hurst (1951) $\log Q_n$ is regressed on $\log n$ in

(4.18) $\log Q_n = a + H \log n + \varepsilon_n$

for various overlapped or non-overlapped samples with n large, and
it is checked whether H is close to 1/2 as $Q_n/\sqrt{n}$ behaves like the
range of a Brownian bridge if $\{x_t\}$ is iid. In Hurst's case the

estimate of H was significantly larger than $1/2$ in terms of t-statistic. Using Taqqu's result (1975), Lo (1991) showed that if $\{x_t\}$ is a Gaussian stationary process with

$$(4.19) \quad \sigma_n{}^2 = \mathrm{Var}(\Sigma_{t=1}^{n} x_t) \sim n^{2H} \quad \text{and}$$

$$(4.20) \quad \gamma_k \sim \begin{array}{ll} k^{2H-2} & H \in (1/2,\ 1) \\ -k^{2H-2} & H \in (0,\ 1/2) \end{array},$$

then Q_n/σ_n converges in distribution to the range of a fractional Brownian bridge on $[0,1]$ defined by

$$V_H = M_H - m_H \quad \text{with} \quad M_H = \max_t FBB_t \quad \text{and} \quad m_H = \min_t FBB_t,$$

where

$$FBB_t = B_t - t\,B_1 \quad \text{with} \quad \{B_t\} \text{ a FBM(H)}.$$

Hence if $\{x_t\}$ is of such a long-term dependency as in (4.19) and (4.20), which is the case of $\{x_t\}$ being a FGN, an estimate of regression coefficient H in (4.18) gives an estimate of H in a FGN(H) model. Lo (1991) proposed another rescaled range statistic adjusted for short-term dependency and used it as a test statistic for the existence of a long-term dependency. According to his result, the null hypothesis that return process $\{x_t\}$ is of no long-term dependency is not rejected for US stocks. Kariya and Katsuura (1992) confirmed the observation for Japanese stocks and Yen/Dollar exchange rates.

Practical option pricing under a FARIMA model

As is observed in Chapter 2, an absolute return series $\{|x_t|\}$, which may be regarded as a volatility series $\{|x_t - \overline{x}|\}$ as $\overline{x}$ is in general small, is of the empirical feature that its autocorrelation are slowly declinig and all positive. This feature may be viewed as a revealed FARIMA feature of $\{x_t\}$ as

$$x_t - \mu = A_t | x_t - \mu | \quad \text{with} \quad A_t = (x_t - \mu)/|x_t - \mu|.$$

If this viewpoint is taken, one may price a call under a FARIMA model in the following procedure.

1) Unknown parameters including d are estimated for a Gaussian FARIMA(p, d, q) model for each given (p, q) by the ML (maximum likelihood) method (see Yajima (1985)).

2) A model is selected for (p, q) probably by such an information criterion (IC) as the Akaike IC or the Schwartz IC.

3) From the selected model with estimated parameters, thousands of time paths are generated to evaluate the mean and variace of a call (contingent claim) as in the case under ARCH model in Section 3.

5 Asian options

Call and put options are standardized options traded in open markets. But in OTC (over-the-counter) markets, various types of options are created according to various demands. Many of them are path-dependent and it is difficult not only to price them theoretically in the manner of Chapter 12 but also to evaluate the expected values of those contingent claims. In this section, as examples, we simply introduce so-called Asian option and discuss about the evaluations of the expected values by a Monte Carlo approach.

Options on the average prices of a price process over a certain period are sometimes called Asian options. A typical one is a call option on the arithemetic average of a price process $\{ S_t \}$ over $[t + u, \ t + v]$ ($v > u$);

$$(5.1) \quad X = \max(\overline{S}_{v-u} - K, 0) \quad \text{with} \quad \overline{S}_{v-u} = \Sigma_{j=u}^{v} S_{t+j}/(v - u),$$

where t is the present time and S_s is the closing price of the s-th day. Hence X is an option on the averaged price process $\{ \overline{S}_{t-u} : t = u, \ u+1, \cdots, \ u + v \}$ and it should be European by its form, i.e., the option can be excercised only on the $(t+v)$-th day. Clearly the value of X depends on how S_{t+j}'s travel over $[t+u,\ t+v]$ and hence it is a path-dependent contingent claim. In this case, even if we can assume that $\{S_t\}$ follows a geometric Brownian motion, it is difficult to evaluate or even efficiently approximate the expected value of this simple but practical option. This is because the domain A of the integral of $E(X)$ is intractable in terms of iid normal variates $x_{j,t+j}$'s or because the distribution of $\overline{S}_{v-u}$ is too complicated. In fact, with $x_{t+j}=\log S_{t+j}-\log S_{t+j-1}$, the domain is expressed as

$$A = \{\Sigma_{h=u+1}^{v}\exp(\Sigma_{j=u+1}^{h}x_{t+j})>\tau K/S_t\},$$

as $S_{t+h}=S_t\exp(\Sigma_{j=1}^{h}x_{t+j})$ and hence the distribution of $\overline{S}_{v-u}$ is hard to approximate. In such a case, as has been discussed in Sections 1 and 3, a suggested procedure is the Monte Carlo evaluation procedure. To apply it, we need to generate thousands or millions of iid normal time paths x_k's where $x_k=(x_{t+u,\ k},\cdots,\ x_{t+v,\ k})$ with each $x_{t+j,\ k}$ randomly generated from $N(\mu,\sigma^2)$. Then for each time path x_k, we compute X_k and obtain $\overline{X}$ as an estimate of $E(X)$. We may also compute the variance of X_k's as an estimate of $\mathrm{Var}(X)$ for a risk measure of the option (see Section 3). In such a Monte Carlo approach, the process of $\{x_t\}$ can be anything as far as its time paths can be generated by computer. Boyle (1991) pointed out that a lognormal approximation to the distribution of $\overline{S}_{v-u}$ will be effective. In his paper, expected values of various options on geometric means of prices are evaluated.

In association with (5.1), the option defined by

(5.2) $X = \max(S_{t+v}-\overline{S}_{v-u},\ 0)$

is also traded. In this option, a buyer of X can gain only if the price S_{t+v} at maturity goes over the average price $\overline{S}_{v-u}$ over $[t+u,\ t+v]$. It is called an average lookback option. Again a Monte

Carlo evaluation for $E(X)$ is needed. We can create many such options but it is difficult to evaluate them analytically.

The difficulty of evaluating $E(X)$ in (5.1) analytically lies in the fact that the distributions of S_t's are not reproductive for addition even under log-normality. Hence it is also difficult to evaluate the expected value of such a call option on multiple assets as

$$(5.3) \quad X_T = \max(\frac{1}{M}\Sigma_{i=1}^{M} S_{iT} - K, 0)$$

where S_{it} is the i-th asset price at t ($i=1,\cdots,M$). If we could assume that S_t's or S_{it}'s are jointly normal, which is sometimes made in the literature, the valuation of $E(X_T)$ or $Var(X_T)$ would be very easy.

Exercises

1. Show (2.1).
2. Verify 1), 2) and 3) for $g(z)$ in (2.2).
3. Give the details of the proof of Lemma 2.1.
4. Empirically observe whether $\overline{x}$ as estimate of u is close to $r - s^2/2$ for your choice of r, where $\{x_t\}$ is a return process and s^2 is sample variance.
5. Verify (2.12).
6. Assuming a given process $\{x_t\}$ is iid, evaluate the Gram-Charlier call premium in (2.7) for your choice of an asset and compare it with the BS value.
7. Verify Lemma 3.1.
8. Verify (3.7), (3.8), (3.9) and (3.11).
9. Verify (3.13) and (3.15).
10. Verify (3.19).
11. Show that a process given by (4.6) is stationary if $|d|<1/2$.

CHAPTER 14

STATISTICAL BOND PRICING MODELS

1 Introduction

In this section, we propose some empirical bond pricing models
by formulating market discount function or spot rate function and
consider their estimation and prediction procedures. Here by
market discount function we mean a function of time and characteristics (attributes) of bonds which prices different bonds consistent-
ly with their different cash flows and different characteristics.
It is a stochastic function. An alternative specification of
duration is also proposed in our model.

The procedure of estimating the market discount function (or
spot rate function) based on market bond prices in general depends
on the formulation of variational structure of bond prices. The
approaches to the formulation of the variational structure are classified as

(1) empirical (statistical) model approach and

(2) normative model approach.

In approach (1), a statistical model is usually specified for the
cash-flow discount function which is common to market and its
estimation and prediction procedure is considered together with how
to make a good use of it for decision making. An example is
McCulloch(1975)'s approach via spline function. In approach (2), a
hypothetical or theoretical world such as a world with no arbitrage
under certainties is assumed and price changes are described with a
stochastic model which is compatible with the conditions derived
under the assumed setting. The approach by Heath, Jarrow and
Morton (1987, 1990) (abbreviated as HJM below) who followed Ho and
Lee (1985) is such an example and they specified a two factor
stochastic model for zero bonds.

In real world uncertainties prevail strongly and bond prices
depend on their characteristics (such as default risk, callability

263

condition, etc.), investment stances of investors, institutional
conditions (tax, trading cost), business conditions, expectations
of investors which will change with change of fundamental surroundings, etc. Hence it is not easy to evaluate the relative advantages
of the two approaches in advance. In fact, it will be more important to test the models empirically, and find out the relative
performances as pricing models for decision-making. And bonds
traded in markets are those with various cash flows and characteristics and most of them are nonzero bonds, bonds with coupons.

2 Basic structure of bond analysis

Here a bond is defined to be a security with a stream of fixed
incomes or cash flows at fixed time points in future. Hence mor-
gage securities are bonds in our terminology. It is assumed that
there are N bonds and t stands for "present" time point. Then
the i-th bond is described at each t as the observable triplet:

(2.1) $(C_{it}(\cdot),\ \mathbb{Z}_{it},\ P_{it}(0))$,

where $C_{it}(\cdot)$ is the cash flow function of the i-th bond, $\mathbb{Z}_{it}$ is
the set of its characteristics and $P_{it}(0)$ is the price of the i-
th bond at t. One may include the past prices and some past
exogenous variables in $\mathbb{Z}_{it}$. But here we do not include them explicitly because we make simply a cross-sectional analysis. The
three elements in (2.1) are explained in details below.

I Cash flow function $C_{it}(s)$; $0 \leq s \leq s_{M(i)}$

Let the time *points* at which cash flows occur be

(2.2) $(t<)\quad t_1 < t_2 < \cdots < t_m < \cdots < t_{M(i)}$

and let the corresponding time periods from the present time t be

(2.3) $s_j = t_j - t \quad (j = 0, \cdots, M(i)),\quad s_0 = 0.$

Here $t_{M(i)} = t + s_{M(i)}$ is the time point of maturity for the i-th bond and hence $s_{M(i)}$ is the *period* till the maturity time point. The cash flows occurring at $t + s_j$ are represented by $C_{it}(s_j)$ ($j = 1, \cdots, M_{(i)}$). But by setting $C_{it}(s) \equiv 0$ except for the time points $t + s_j$ ($j = 1, \cdots, M_{(i)}$) we can regard the cash flow function $C_{it}(s)$ as a function defined on $[0, s_{M(i)}]$;

$$(2.4) \quad C_{it}: [0, s_{M(i)}] \quad \rightarrow \quad [0, \infty).$$

The function $C_{it}(\cdot)$ is known at each time t.

‖ **Bond characteristics** $Z_{it} = \{z_{ikt}: k = 1, \cdots, q\}$

For each bond i, it is assumed that there are q observable, non-stochastic and exogenous (or predetermined) variables $\{z_{ikt}: k = 1, \cdots, q\}$ at t. For example, z_{i1t} represents the kind of bonds, z_{i2t} the rating for default risk, z_{i3t} the call condition, etc.

‖ **Price** $P_{it}(0)$ at t

$P_{it}(s)$ denotes the future price of time point $t + s$ viewed at time t ($s > 0$) and $P_{it}(0)$ is the price of present time t. Only the price $P_{it}(0)$ at t is an observable random variable and prices $P_{it}(s)$ with $s > 0$ are unobservable random variables, forming a stochastic process at each t for $0 \leq s \leq s_{M(i)}$;

$$(2.5) \quad P_{it} = \{P_{it}(s): 0 \leq s \leq s_{M(i)}\}, \quad P_{it}(s_{M(i)}) = 100.$$

This process changes with t, which corresponds to the changes of the term structure embedded in the bond price process. At maturity $t + s_{M(i)}$ $P_{it}(s_{M(i)}) = 100$ with probability 1. The reason why we consider nonobservable $P_{it}(s)$ ($s > 0$) is that investors have to forecast future price $P_{it}(s)$ at t with given information when they buy at t and plan to sell at $t + s$. However no model for P_{it} is given in this chapter.

3 Market spot rate function

In most cases, spot rate is defined to be the discount rate evaluating the present value of zero (coupon) bond and given as

(3.1) $d_{it}(s_{M(i)}) = -[\log P_{it}(0) - \log P_{it}(s_{M(i)})]/s_{M(i)}.$

Since $P_{it}(s_{M(i)}) = 100$, the stochastic variational structure of $P_{it}(0)$ corresponds to that of $d_{it}(s_{M(i)})$. In fact (3.1) is equivalent to

(3.2) $P_{it}(0) = 100\, D_{it}(s_{M(i)}),$ where

(3.3) $D_{it}(s) = \exp(-s\, d_{it}(s))$ and $D_{it}(s_{M(i)}) = P_{it}(0)/100.$

Hence $d_{it}(s_{M(i)})$ in (3.1) is simply a one-to-one transformation of price $P_{it}(0)$. However we regard d_{it} as a spot rate function defined on $[0, s_M]$

(3.4) $d_{it} : [0, s_M] \to [0,1]$, where $M = \max\{M(1), \cdots, M(N)\},$

and as in (3.2) regard the price $P_{it}(0)$ as the value discounted by the discount function $D_{it}(s)$ with $s = s_{M(i)}$ associated with $d_{it}(s)$ through (3.3). In this viewpoint we make the following remarks.

i) A discount function $D_{it}(s)$ forms a stochastic process

(3.5) $\boldsymbol{D}_{it} = \{D_{it}(s) : 0 \le s \le s_M\}$

and specifying $\boldsymbol{D}_{it}$ is equivalent to giving a pricing model for $P_{it}(0)$. In practice, it is often the case that $\boldsymbol{D}_{it}$ is regarded as the causal variation of $P_{it}(0)$. Spot rate function $d_{it}(s)$ is a one-to-one function of $D_{it}(s)$ via (3.3).

ii) In general, D_{it} (or d_{it}) depends on bond characteristics such as callability condition, default risk, etc. On the other hand, the discount function $D_{it}(s)$ of the i-th bond is observable through (3.3) only for $s = s_{M(i)}$ at t.

Hence in order to secure the statistical estimability of the

functional form D_{it} (or d_{it}), which may be stochastic, from data
$D_{it}(s_{M(i)})$'s (or $d_{it}(s_{M(i)})$'s) ($i=1,\cdots,N$), a certain ex ante
model is required.   Such a model will be classified as (1) statistical model or (2) normative model.

To extend the above argument on zero bonds to the case of coupon
bonds, we consider the two types of cash-flow discount functions;
(1) subjective discount (or spot rate) function,
(2) market discount (or spot rate) function.

(1) Subjective discount function

Suppose an investor has investment period $(0, s_L]$ with s_L less
than maturity $s_{M(i)}$ and plans to purchase the i-th bond.   Then
cash flows from the investment are generated at s_j's ($j=1,\cdots,m$)
with $s_m \leq s_L < s_{m+1}$ ($< s_{M(i)}$).   In this situation,
  i) what is certain at t is ($C_{it}(\cdot)$, $\mathbf{Z}_{it}$, $P_{it}(0)$), and
  ii) what is uncertain at t is price $P_{it}(s_L)$ at t.
Since investment time horizon is $(0, s_L]$, the cash flows $\{C_{it}(s_j)\}$
obtainable after periods $s_1, \cdots, s_m$ can be reinvested till $t + s_L$
time point.   Let $b_{it}(s)$ denote the reinvestment rate at $t + s$
subjectively forecasted at t.   The forecast naturally involves uncertainty.   The compound interest rate function based on $b_{it}(s)$ is
denoted by

(3.6) $B_{it}(s, s_L)=\exp((s_L - s) b_{it}(s))$, $0 < s \leq s_L$.

Then the cash flow $C_{it}(s_j)$ obtainable at $t + s_j$ and reinvested
at forecasted rate $b_{it}(s_j)$ till $t + s_L$ is expected at t to grow
up to $C_{it}(s_j) B_{it}(s_j, s_L)$; $C_{it}(s_j) \to C_{it}(s_j) B_{it}(s_j, s_L)$.
Hence revenue at selling time $t + s_L$ is forecasted at t to be

(3.7) $\sum_{j=1}^{m} C_{it}(s_j) B_{it}(s_j, s_L) + P_{it}(s_L)$.

To discount this revenue to the present value, let a subjective
spot rate function be $\delta_{it} : [0, s_L] \to [0,1]$ and let its discount
function be $\Delta_{it}(s, s_L)=\exp(- s_L \delta_{it}(s))$ which discounts

forecasted income $C_{it}(s) B_{it}(s, s_L)$ at $t + s_L$ resulting from
cash flow $C_{it}(s)$ reinvested at the rate $b_{it}(s)$ at $t + s$. Then
the expected present value of $C_{it}(s) B_{it}(s, s_L)$ is $C_{it}(s) \widetilde{D}_{it}(s)$
where

$$(3.8) \qquad \widetilde{D}_{it}(s) \equiv \widetilde{D}_{it}(s, s_L) = B_{it}(s, s_L) \varDelta_{it}(s, s_L)$$
$$= \exp((s_L - s) b_{it}(s) - s_L \delta_{it}(s)).$$

Therefore the total forecasted income is subjectively evaluated as

$$(3.9) \qquad Q_{it}(s_L) = \sum_{j=1}^{m} C_{it}(s_j) \widetilde{D}_{it}(s_j) + P_{it}(s_L) \widetilde{D}_{it}(s_L).$$

Here the reasons why the values of discount function $\varDelta_{it}(s_j, s_L)$
are different depending on the reinvestment times $t + s_j$'s though
all the incomes are commonly generated at s_L are:
1) the reinvestment rates are based on prediction and the reliability of reinvestment rates in distant future is low, and
2) the reinvestment rates greatly depend on business surroundings
   in future, affecting general interest levels which need to be
   forecasted.
This implies that the subjective spot rate function $\delta_{it}(s)$ is formed through the subjective distribution which depends on the forecasts of future business conditions. For the exponent part in
$\widetilde{D}_{it}(s)$ of (3.8); $-s_L[\delta_{it}(s) - b_{it}(s)] - s b_{it}(s)$, it is usually
assumed that (i) $\delta_{it}(s) = b_{it}(s)$ in which case

$$(3.10) \qquad \widetilde{D}_{it}(s) = \exp(- s \delta_{it}(s)).$$

However it is likely that individual investors may have (ii) $\delta_{it}(s)$
$\neq b_{it}(s)$. In the sequal, we simply consider the customarily treated case (i).

    Even in the case (i), it is noted that the present value of the
total forecasted income $Q_{it}(s_L)$ in (3.9) is based on the prediction. In fact,
  a) subjective reinvestment rates are forecasted values, and

b) $P_{it}(s_L)$ is a predicted value.

In other words, uncertainties involved in the evaluation of Q_{it} (s_L) are multifold. Giving the subjective forecasted values of $\delta_{it}(s_j)$ and $P_{it}(s_L)$, one may evaluate $Q_{it}(s_L)$ and compare it with the present market value $P_{it}(0)$ to make an investment decision. However in actual decision-making prediction errors (risks) involved in a) and b) need to be taken into account. Hence it will be better to treat $\delta_{it}(s)$ and $P_{it}(s_L)$ as random variables, which is the next theme.

(2) Market discount function

The above argument is here developed to formulate a pricing model in the whole market. First note that there are many investors and they have different time horizons with different risk preferences and different prediction procedures. Hence it is natural to presume in (3.9) the subjective functions $\delta_{it}(s)$ and $\widetilde{D}_{it}(s)$ are replaced by random market functions $d_{it}(s)$ and $D_{it}(s)$. Hence the present market price $P_{it}(0)$ of the i-th bond is randomly determined by $P_{it}(0) \equiv Q_{it}(s_{M(i)})$ with $\widetilde{D}_{it}(s) = D_{it}(s)$;

$$(3.11) \quad P_{it}(0) \equiv \Sigma_{j=1}^{M(i)} C_{it}(s_j) D_{it}(s_j) + 100 D_{it}(s_{M(i)})$$
$$\text{with} \quad D_{it}(s) = \exp(-s d_{it}(s)).$$

Here $D_{it}(s)$ is a random variable for each s, and hence $\boldsymbol{D}_{it} = \{D_{it}(s) : 0 \leq s \leq s_{M(i)}\}$ is a stochastic process. Thus the price model $(C_{it}(\cdot), \boldsymbol{Z}_{it}, P_{it}(0))$ in (2.1) is viewed as

$$(3.12) \quad (C_{it}(\cdot), \boldsymbol{Z}_{it}, \boldsymbol{D}_{it}).$$

In this viewpoint the randomness of $P_{it}(0)$ is inherited from $\boldsymbol{D}_{it}$. However, we confront the problem of the identifiability of the process $\boldsymbol{D}_{it}$. In other words, for given $C_{it}(\cdot)$ and $P_{it}(0)$ ($i=1, \cdots, N$), there are many discount functions which satisfy (3.11). Hence we need to specify the process $\boldsymbol{D}_{it}$ in advance so that from cross-sectional data $(C_{it}(\cdot), \boldsymbol{Z}_{it}, P_{it}(0))$ ($i=1, \cdots, N$), $\boldsymbol{D}_{it}$

can be identified and estimated. In zero coupon bonds, $P_{it}(0)$ and $d_{it}(s_{M(i)})$ are in one-one correspondence and hence $D_{it}(s_{M(i)})$ is obtained uniquely from $P_{it}(0)$. Note that although $D_{it}(\cdot)$ carries the suffix i associated with characteristis $\mathbf{Z}_{it}$, there are some common (random) factors and parameters in $D_{it}(\cdot)$ affecting all the bonds in market.

So far we discussed the problem for fixed t (present time). What is more important from a viewpoint of investment will be how prices $\{P_{it}(0)\}$ evolve over time. At time t with available information we may suppose an imaginary evolutionary process $\mathbf{P}_{it} = \{P_{it}(s): 0 \leq s \leq s_{M(i)}\}$, which terminates at maturity $t + s_{M(i)}$, where we know $P_{it}(s_{M(i)}) = 100$ with probability one. However, at $t + t_1$ $P_{it}(t_1)$ we supposed at t is determined by the different process $\mathbf{P}_{it+t_1}$ and it is different from $P_{it+t_1}(0) \in \mathbf{P}_{it+t_1}$, which is caused by the change of the process $\mathbf{D}_{it}$ of discount function at t into $\mathbf{D}_{it+t_1}$. In investment decision-making it is important to make a bridge between $\mathbf{D}_{it}$ and $\mathbf{D}_{it+t_1}$ so that we can predict in what way the term structure of interest at t is deformed at $t + t_1$. We discuss this problem in Section 6.

4 Market spot rate function of zero coupon bond

When we view the variational structure of the i-th bond at time point t as $(C_{it}(\cdot), \mathbf{Z}_{it}, \mathbf{D}_{it})$, as we discussed in Section 3, we need to specify the stochastic process of the discount function $D_{it}(s)$ on $0 \leq s \leq s_{M(i)}$ since

$$(4.1) \quad P_{it}(0) = \Sigma_{j=1}^{M(i)} C_{it}(s_j) D_{it}(s_j),$$

where $C_{it}(s_{M(i)})$ here denotes $C_{it}(s_{M(i)}) + 100$ in Section 3. It is noted that $D_{it}(s_j)$'s $(j=1,\cdots,M(i))$ are correlated random variables, may depend on the characteristics $\mathbf{Z}_{it}$ of each bond and are correlated with $D_{kt}(s_j)$'s of other bonds. In specifying $\mathbf{D}_{it}$, the following points need to be taken into account;

1) heteroscedastic property of $P_{it}(0)$ that when the period $s_{M(i)}$
   till maturity is short, the variance of $P_{it}(0)$ is small,
2) how to separate common market variations from variations which
   are specific to each bond characteristics,
3) correlations among different processes D_{it}'s ($i=1,\cdots,N$), and
4) parsimonous parametrization.

The specification problem is treated below. In this section
statistical models for zero bonds are proposed and in the next section, statistical models for coupon bonds (including zero bonds)
are proposed.

Let $M=\max M(i)$ and regard $D_{it}(s)$ as a stochastic process on
$[0, s_M]$. In the case of zero bonds, it holds that

$$P_{it}(0)=100\, D_{it}(s_{M(i)})=100\ \exp(- s_{M(i)}\, d_{it}(s_{M(i)})).$$

Hence we may specify either the random spot rate function $d_{it}(s)$
or the discount function $D_{it}(s)$. But they are not the same
because

$$E[D_{it}(s)] \neq \exp(- s\, E[d_{it}(s)]).$$

(1) A model for market spot rate function

As a simple model, we consider an additive model;

(4.2) $d_{it}(s)= \mu_t(s)+\gamma_{it}(s)+ \varepsilon_{it}(s)$,   where

     i) $\mu_t(s)=\delta_{0t}+\delta_{1t}\, s+\delta_{2t}\, s^2+ \cdots + \delta_{pt}\, s^p$

(4.3) ii) $\gamma_{it}(s)=\alpha_{1t}(s)\, z_{i1t}+ \cdots + \alpha_{1qt}(s)\, z_{iqt}$

    iii) $\varepsilon_{it}(s)$'s are random error terms with $E[\varepsilon_{it}(s)]=0$
         and $\sigma_{ikt}(s, s')=\mathrm{Cov}(\varepsilon_{it}(s), \varepsilon_{ik}(s'))$.

As $\sigma_{ikt}(s, s')$ in view of 1)~4) above, we consider the form

(4.4) $\sigma_{ikt}(s, s') = \lambda_{ikt}\min(c_i(s), c_k(s'))$,

where $\Lambda=(\lambda_{kit})$ is specified later and $c_i(s)$'s are known function such that $c_i(0)=0$ and $c_i(s)>0$ for $s>0$. In (4.2), the mean of $d_{it}(s)$ is decomposed into the market common part $\mu_t(s)$ which does not depend on the characteristics $\mathbf{Z}_{it}$ of each bond and the part $\gamma_{it}(s)$ which depends on the characteristics $\mathbf{Z}_{it}$. Further it is assumed that $\mu_t(s)$ is approximated by a polynomial of s. Of course it may be specified as a spline function as in McCulloch (1975). Also $\gamma_{it}(s)$'s are assumed to be approximated by a linear function of z_{ijt}'s. We also assume that the coefficients $\alpha_{jt}(s)$ are approximated by simple functions of s. In particular, we assume for simplicity

(4.5) $\alpha_{jt}(s) \equiv \alpha_{jt}.$

The specification of the stochastic structure of error terms $\varepsilon_{it}(s)$ is that of the N-dimensional multivariate stochastic process for

(4.6) $\varepsilon_t(s)=(\varepsilon_{1t}(s), \cdots, \varepsilon_{Nt}(s))',   0 \leqq s \leqq M.$

But the observations are available only for

(4.7) $s_{M(1)} \leqq s_{M(2)} \leqq \cdots \leqq s_{M(N)},$

and hence as in (4.3) iii), it is sufficient to specify the probabilistic structure of the N random variables

(4.8) $(\varepsilon_{1t}(s_{M(1)}), \cdots, \varepsilon_{Nt}(s_{M(N)})).$

In (4.4) a heteroscedastic nature of variations is taken into account as stated in 1) above and a monotone increasing function of s will be promising for $c_i(s)$. As a specific choice of $c_i(s)$, we assume in the sequal

(4.9) $c_i(s) = s_{M(i)}   (i=1, \cdots, N).$

We still need to specify λ_{ikt}'s supposedly representing correla-

tions of $\varepsilon_{it}(s_{M(i)})$'s.   Otherwise too many parameters are involved in the model relative to N observations.   As a candidate for
λ_{ikt}'s, we propose

$$(4.10) \quad \lambda_{ikt} = \frac{\sigma^2 a_{iit}}{\sigma^2 \rho \, a_{ikt}} \, (i \neq k), \quad \text{where}$$

$$(4.11) \quad a_{ikt} = b(z_{it}, z_{kt})\exp(-|s_{M(i)} - s_{M(k)}|),$$

and $z_{it}=(z_{i1t}, \cdots, z_{iqt})'$ with $b(z_{it}, z_{kt})$ a known function of
z_{it} and z_{kt}.   It is noted that $d_{it}(s)$ can be formulated as (4.2)
only for zero bonds where $P_{it}(0)$ and $d_{it}(s_{M(i)})$ are uniquely connected.

Estimation

The model (4.2) is expressed as a regression model;

$$(4.12) \quad y_t = X_t \beta_t + \eta_t \quad \text{where}$$

(4.13)

$$y_t = (d_{1t}(s_{M(1)})/s_{M(1)}^{1/2}, \quad \cdots \quad , d_{Nt}(s_{M(N)})/s_{M(N)}^{1/2})' : N \times 1,$$
$$X_t = (x_{ijt}) : N \times (p+q+1) \, (i=1, \cdots, N : j=0,1, \cdots, p+q),$$
$$x_{i0t} = s_{M(i)}^{-1/2}, \quad x_{ijt} = s_{M(i)}^{\,j}/s_{M(i)}^{1/2} \, (j=1, \cdots, p),$$
$$x_{ijt} = z_{ij-pt}/s_{M(i)}^{1/2} \, (j=p+1, \cdots, p+q),$$
$$\beta_t = (\delta_{0t}, \delta_{1t}, \cdots, \delta_{pt}, \alpha_{1t}, \cdots, \alpha_{qt})' : (p+q+1) \times 1, \quad \text{and}$$
$$\eta_t = (\varepsilon_{1t}(s_{M(1)})/s_{M(1)}^{1/2}, \quad \cdots \quad , \varepsilon_{Nt}(s_{M(N)})/s_{M(N)}^{1/2})',$$

In this expression, the covariance matrix of η_t is

$$(4.14) \quad \text{Cov}(\eta_t) = \Lambda_t = \sigma^2[(1-\rho)J_t + \rho A_t] \equiv \sigma^2 \Lambda_{0t}(\rho),$$

where $J_t = \text{diag}\{a_{11t}, \cdots, a_{NNt}\}$ and $A_t = (a_{ijt})$.   Hence by the GLS
(generalized least squares) method we minimize

$$(4.15) \quad \phi(\rho, \beta_t) = (y_t - X_t \beta_t)' \Lambda_{0t}(\rho)^{-1} (y_t - X_t \beta_t)$$

possibly via the Newton-Raphson method (or a repeated method: $\rho(0)$

$=0 \to \min\phi_t(0, \beta_t) \to \beta_t(1) \to \min\phi_t(\rho, \beta_t(1)) \to \rho(1) \to \cdots)$.
Then the GLSE (GLS estimator) is $(\hat{\beta}_t, \hat{\rho})$ with

(4.16) $\hat{\beta}_t = (X_t' \Lambda_{0t}(\hat{\rho})^{-1} X_t)^{-1} X_t' \Lambda_{0t}(\hat{\rho})^{-1} y_t$,

from which $\hat{\delta}_{jt}$ and $\hat{\alpha}_{jt}$ are obtained. For this estimator to be
effective, it is desired that
 i) N is large and
 ii) $s_{M(i)}$'s are spread out.
But in case of zero bonds these conditions are unlikely to be satisfied.

It is remarked that in place of the additive specification of
μ_t and γ_{it} in (4.2) it may be specified as

$d_{it}(s) = \mu_{it}(s) + \varepsilon_{it}(s)$ with $\varepsilon_{it}(s)$ in (4.3) iii),
$\mu_{it}(s) = \delta_{0t}(z_{it}) + \delta_{1t}(z_{it}) s + \cdots + \delta_{pt}(z_{it}) s^p$ and
$\delta_{jt}(z_{it}) = \alpha_{0t} + \alpha_{1t} z_{i1t} + \cdots + \alpha_{qt} z_{iqt}$.

This specification also leads to a regression model. One may also
attempt to estimate $\sigma_{ikt}(s, s')$ directly by pooling time series
data under the assumption that $\sigma_{ikt}(s, s') = \sigma_{ik}(s, s')$ for all
t.

(2) Modelling discount function $D_{it}(s)$

Another specification for modelling $P_{it}(0)$ in case of zero
bonds is to model $D_{it}(s)$ directly as $D_{it}(s) = \exp(-s\, d_{it}(s))$.
This is because even under normality assumption for $\varepsilon_{it}(s)$ in (4.2)
the expected value of $P_{it}(0)$ incurrs the effect of the variance
$\sigma_{iit}(s, s)$ of $\varepsilon_{it}(s)$ when $d_{it}(s)$ is specified. In fact,

(4.17) $E[P_{it}(0)] = 100\, E[D_{it}(s)]$
$= 100\exp[-s\mu_t(s) - s\gamma_{it}(s) + \frac{1}{2} s^2 \sigma_{iit}(s, s)]$

with $s = s_{M(i)}$. There may be some cases in which the effect due
to $s^2 \sigma_{iit}(s, s)$ is desirable. But the specific effect of each

bond is also considered doubly counted in view of the role of γ_{it} (s) and it may be very big for s large. Hence one may separate the mean part of $D_{it}(s)$ from $D_{it}(s)$ as

(4.18) $P_{it}(0) = 100 E[D_{it}(s_{M(i)})] + \eta_{it}$ with
$\eta_{it} = 100\{D_{it}(s_{M(i)}) - E[D_{it}(s_{M(i)})]\}$

and then specify the mean part and the error part separately. This approach is valid for coupon bonds as well and hence it is discussed in the next section.

5 Statistical model for market discount function of coupon bonds

Let N be the number of available bonds in the market (coupon, zero coupon or other fixed income securities). The time points when cash flows are generated are in general different in these bonds. Hence let

(5.1) $s_j = s(i)_j$ $(j=1, \cdots, M(i): i=1, \cdots, N)$

denote the periods from present time t to the time the cash flows of the i-th bond occur. Let the enumeration of all these periods in the ascending order be

(5.2) $s_{a1} < s_{a2} < \cdots < s_{aM}$, $M = \max M(i)$,

and regard $C_{it}(s)$ as a function defined on $0 \leq s \leq s_{aM}$, where $P_{it}(s_{M(i)}) = 100$ is included in $C_{it}(s)$. The model we consider in the sequal is

(5.3) $P_{it}(0) = \sum_{j=1}^{M} C_{it}(s_{aj}) D_{it}(s_{aj}) = C_{it}' D_{it}$

with $C_{it} = (C_{it}(s_{a1}), \cdots, C_{it}(s_{aM}))'$ and $D_{it} = (D_{it}(s_{a1}), \cdots, D_{it}(s_{aM}))'$, where D_{it} is a random vector. Note unless $s_{ak} = s(i)_j$, $C_{it}(s_{ak}) = 0$.

(1) Specification of D_{it}

As discussed in Section 4, $D_{it}(s)$ is directly modelled by separating the mean part and the error part in (5.3) as

(5.4) $P_{it}(0) = C_{it}' \bar{D}_{it} + \eta_{it}$, with $\bar{D}_{it} = E(D_{it})$,
            $\eta_{it} = C_{it}' \nu_{it}$ and $\nu_{it} = D_{it} - \bar{D}_{it}$.

As specifications of $\bar{D}_{it}(s)$, we may take spline functions or polynomials as in McCullosh (1971) or in Section 4 respectively. In such cases the basic structure of the argument is the same as the one below. In any case the problem is how to incorporate the characteristics of each bond into the model. In this section we consider the case where the coefficients in a polynomial depend on the characteristics:

(5.5) $D_{it}(s) = 1 + \delta_{1t}(z_{it}) s + \cdots + \delta_{pt}(z_{it}) s^p$,   where

            $\delta_{jt}(z_{it}) = \delta_{j0t} + \delta_{j1t} z_{i1t} + \cdots + \delta_{jqt} z_{iqt}$,
            $\mathrm{Cov}(\eta_{it}, \eta_{kt}) = \lambda_{ikt} C_{it}' \Phi_{ikt} C_{kt}$, and
            $\Phi_{ikt} = \mathrm{Cov}(\nu_{it}, \nu_{kt})$.

Let $\Phi_{ikt} = (\phi_{ikt \cdot jr})$. Then $\phi_{ikt \cdot jr}$ is the covariance between $D_{it}(s_{aj})$ and $D_{kt}(s_{ar})$, and for example it is specified as

(5.6) $\phi_{ikt \cdot jr} = f(s_{aj}, s_{ar}) = \exp(-|s_{aj} - s_{ar}|)$.

In this specification, let

(5.7) $f_{ikt} = C_{it}' \Phi_{ikt} C_{kt}$ and $F_t = (f_{ikt})$.

Also let us use the specification (4.11) for λ_{ikt}. Then the model is again expressed as a regression model:

(5.8) $y_t = X_t \beta_t + \eta_t$,   where

            $y_t = (y_{it})$ with $y_{it} = P_{it}(0) - \sum_{j=1}^{M} C_{it}(s_{aj})$,
            $\beta_t = (\delta_{1t}', \cdots, \delta_{pt}')'$ with $\delta_{kt} = (\delta_{k0t}, \cdots, \delta_{kqt})'$,

$X_t = (x_{ijt})$ with $x_{ijt} = \sum_{k=0}^{P} z_{ikt} s_{aj}^k C_{it}(s_{aj})$,

$\eta_t = (\eta_{1t}, \cdots, \eta_{Nt})'$, $\mathrm{Cov}(\eta_t) = \sigma^2 \Sigma_t = \sigma^2(\sigma_{ijt})$,

$\sigma_{iit} = a_{iit} f_{iit}$, and $\sigma_{ijt} = \rho\, a_{ijt} f_{ijt}$.

Here $z_{i0t} = 1$. Hence minimizing

(5.9) $(y_t - X_t\beta_t)' \Sigma_t(\rho)^{-1}(y_t - X_t\beta_t)$

with respect to β_t and ρ yields the GLSE of β_t:

(5.10) $\hat{\beta}_t = (X_t' \Sigma_t(\hat{\rho})^{-1} X_t)^{-1} X_t' \Sigma_t(\hat{\rho})^{-1} y_t$.

(2) Modelling market spot rate function

In $D_{it}(s) = \exp(-s\, d_{it}(s))$, we model $d_{it}(s)$. As a process for $d_{it}(s)$, assume (4.2) i) ii) iii) with

(5.11) $\varepsilon_{it}(s) \sim N(0, \sigma_{iit}(s, s'))$ and

$\mathrm{Cov}(\varepsilon_{it}(s), \varepsilon_{kt}(s')) = \sigma_{ikt}(s, s') = \sigma_{ikt}(s', s)$.

Here for $\sigma_{ikt}(s, s')$, for example we take

(5.12) $\sigma_{ikt}(s, s') = \lambda_{ikt} \exp(-|s - s'|)$.

In that case

(5.13) $E[D_{it}(s)] = \exp(-s\, \bar{d}_{it}(s) + \frac{1}{2} s^2 \sigma_{iit}(s, s))$.

Hence letting $h_{it}(s, \theta_t) = E[D_{it}(s)]$ and

$h_{it}(\theta_t) = (h_{it}(s_{a1}, \theta_t), \cdots, h_{it}(s_{aM}, \theta_t))'$,

we obtain $E(y_{it}) = C_{it}' h_{it}(\theta_t)$, where θ_t is a vector of all the parameters involved in the model. Also since

(5.15) $\phi_{ikt \cdot jr} = \mathrm{Cov}(D_{it}(s_{aj}), D_{kt}(s_{ar}))$

$= g_{ik}(s_{aj}, s_{ar}, \theta_t) h_{it}(s_{aj}, \theta_t) h_{kt}(s_{ar}, \theta_t)$

with $g_{ik}(s_{aj}, s_{ar}, \theta_t) = \exp(s_{aj} s_{ar} \sigma_{ikt}(s_{aj}, s_{ar})) - 1$, letting

$\Phi_{ikt} = (\phi_{ikt \cdot jr})$, we obtain

$$\mathrm{Cov}(y_{it}, \, y_{kt}) = C_{it}' \, \Phi_{ikt} C_{kt} \equiv \omega_{ikt}(\theta_t).$$

Hence setting $\tau_t(\theta) = (h_{1t}(\theta_t), \cdots, h_{Nt}(\theta_t))'$ and $\Omega_t(\theta_t) = (\omega_{ikt}(\theta_t))$, and minimizing $(y_t - \tau_t(\theta_t))' \, \Omega_t(\theta_t)^{-1}(y_t - \tau_t(\theta_t))$ with respect to θ_t, we obtain an estimate of the mean of market discount function in (5.13) and a model value of each bond price in (5.14).

6 Prediction of market discount function

In any case in Sections 3, 4 and 5, market discount functions (or market spot rate functions) with structural parameter θ_t are estimated based on cross-sectional data for each time t. The estimate $\hat{\theta}_t$ is a function of z_{it}'s and $P_{it}(0)$'s ($i = 1, \cdots, N$). Hence for predicting future market discount function, it is suffi- cient to predict future θ_t's. As such a method, we may model $\hat{\theta}_t$, $t = 1, \cdots, T$ by the MTV (multivariate time series variance component) model in Chapter 5 or the VAR (vector autoregressive) model in Chap- ter 4, and we can predict future discount functions and hence future bond prices.

On the other hand, as we have pointed out, there are many ways to specify the stochastic process D_{it} of market discount function and data has no power to identify the process.

7 Duration in our statistical models

In the model (5.4) with (5.5) for coupon bonds, we consider the evolutions of the term structure as time t passes. When time t passes to $t + h$, the price of the i-th bond at $t + h$ is expres- sed in our model as

(7.1) $\quad P_{it+h}(0) = \Sigma_{j=i}^{M(i)} C_{it+h}(s_j - h) \overline{D}_{it+h}(s_j - h) + \eta_{it+h}.$

At time $t+h$ the time points at which the cash flows occur are unchanged as $t+h+s_j-h=t+s_j$ though the periods from $t+h$ till the time points are reduced to s_j-h ($j=1,\cdots,M(i)$). Hence it holds that

(7.2) $C_{i\,t+h}(s_j-h) = C_{i\,t}(s_j)$, $j=1,\cdots,M(i)$.

In the time passing the changes of the variational structure from $P_{i\,t}(0)$ to $P_{i\,t+h}(0)$ consist of

(1) the change of the mean market discount function from $\overline{D}_{i\,t}(s)$ to $\overline{D}_{i\,t+h}(s)$, and

(2) the change of the stochastic structure of error term from $\eta_{\,i\,t}$ to $\eta_{\,i\,t+h}$.

In the sequal we discuss these changes separately.

(1) Bond Ψ_t

The reason why we have treated separately starting time t for investment and periods s for term structure so far is that the term structure as a function of s is defined at each time point t and its changes are treated as a function of t in terms of the change of the mean discount function $\overline{D}_{i\,t}(\cdot)$. Here to treat the change with respect to t, let

(7.3) $G(t,s) = \overline{D}_{i\,t}(s)$ $(t,s)\in[0,\infty)\times[0,\infty)$,

and assume that it is expressed as a Taylor expansion;

(7.4) $G(t+h,\ s+k)= G(t,s)+ G_t(t,s)h+ G_s(t,s)k$
$\quad + \frac{1}{2}[\,G_{tt}(t,s)h^2+2G_{ts}(t,s)hk+ G_{ss}(t,s)k^2]+\cdots$,

where G_t, G_{ts}, etc. are partial derivatives with respect to each suffix. Hence the change of $\overline{D}_{i\,t}(s)$ due to the change in time is given by taking $k=-h$;

(7.5) $\overline{D}_{i\,t+h}(s-h)=\overline{D}_{i\,t}(s)+ \phi_{\,i\,t}(s)h+\frac{1}{2}\phi_{\,i\,t}(s)h^2+\cdots$

where $\phi_{it}(s) = \frac{\partial}{\partial t}\overline{D}_{it}(s) - \frac{\partial}{\partial s}\overline{D}_{it}(s)$ and

$$\phi_{it}(s) = \frac{\partial^2}{\partial t^2}\overline{D}_{it}(s) - 2\frac{\partial^2}{\partial t \partial s}\overline{D}_{it}(s) + \frac{\partial^2}{\partial s^2}\overline{D}_{it}(s).$$

In the case of usual bond duration, the change of the functional form of $D_{it}(\cdot)$ with respect t is assumed to be small, and the parallel shift of the spot rate curve is only considered. However, the discount function or its corresponding spot rate function (term structure) is very likely to change over time. Hence the expression in (7.5) is more practical as well as more general. In such an expression, the convexity of spot rate curves does not hold and hence the analysis usually taken is not effective as it stands. In addition, the stochastic nature of the discount function is not taken into account in the usual analysis.

To express the price change relative to the change of the mean discount function, let $\overline{P}_{it}(0) = E[P_{it}(0)]$. Then from (7.2) and (7.5)

$$\overline{P}_{it+h}(0) = \overline{P}_{it}(0) + \Sigma_{j=1}^{M(i)} C_{it}(s_j)\phi_{it}(s_j) h + o(h)$$

where $\lim_{h\to0} o(h)/h = 0$. Hence in the time passing $t \to t+h$ the mean price change due to the change of the mean market discount function is given by

(7.6) $d\overline{P}_{it}(0)/dt = \Sigma_{j=1}^{M(i)} C_{it}(s_j)\phi_{it}(s_{jt})$

This is the change of the mean price due to the change of the mean market discount function equipped with the term structure with in, and it corresponds to the price change $dP_{it}(0)/dr$ due to a parallel shift usually considered. As a predictive modified duration, we may define

(7.7) $\Psi_{it} = (d\overline{P}_{it}(0)/dt)/\overline{P}_{it}(0).$

Then this quantity measures the relative price change due to the change of term structure through time passing and serves as a measure for bond investment. We call it bond "pusi". To evaluate it explicitly, it is required to make an explicit specification of the mean market discount function $\overline{D}_{it}(s)$. To demonstrate an example, let

$$\overline{D}_{it}(s) = \exp(-s\,d_t(s)), \text{ and } d_t(s) = \delta_{0t} + \delta_{1t}s + \delta_{2t}s^2.$$

Then $\psi_{it}(s)$ is evaluated as

$$\psi_{it}(s) = \overline{D}_{it}(s)\left[-s\frac{\partial}{\partial t}d_t(s) + \delta_{0t} + 2\delta_{1t}s + 3\delta_{2t}s^2\right].$$

Hence if δ_{jt}'s are estimated and δ_{jt+h}'s are predicted at t, then the changes of δ_{jt}'s are obtained as

$$d\delta_{jt}/dt \approx (\delta_{jt+h} - \delta_{jt})/h = \gamma_{jt}.$$

Thus we obtain

$$\psi_{it}(s) = \overline{D}_{it}(s)\left[\delta_{0t} + (\delta_{1t} - \gamma_{1t})s + (2\delta_{2t} - \gamma_{1t})s^2 - \gamma_{3t}s^3\right]$$

and hence Ψ_{it} by (7.6) and (7.7).

(2) Changes of η_{it}

The change $\eta_{it} \to \eta_{it+h}$ of error term $\overline{\eta}_{it} = C_{it}'[D_{it} - \overline{D}_{it}]$ through time passing is not directly evaluated because it is a random variable. But a method to measure the size of the change will be to consider the standard deviation of price change $P_{it+h}(0) - P_{it}(0)$. In other words, we may evaluate the variance

(7.8) $\quad \theta_t(h) = \text{Var}(P_{it+h}(0) - P_{it}(0)) = \text{Var}(\eta_{it+h} - \eta_{it})$

and evaluate the reliability of Ψ_{it}. If $\theta_t(h)$ is quite large, then a decision based on Ψ_{it} is subject to a big uncertainty. The evaluation of $\theta_t(h)$ in (7.8) needs a specification of the time series process for η_{it} with respect to t. If they are uncorrelated, then $\theta_t(h) = \mathrm{Var}(\eta_{i\,t+h}) + \mathrm{Var}(\eta_{it})$ which can be evaluated based on the specification of the stochastic structure of η_{it} in Sections 4 and 5. This problem of specifying the time series structure for η_{it} will be left to the readers.

APPENDIX

Chapter 2

1 Nonlinearity implies nonnormality

(1) Let $\{x_t\}$ be a stationary process with $E(x_t)=0$. It is known that $\{x_t\}$ has a spectral density $h(\omega)$ if and only if

(1.1) $\quad x_t = \Sigma_{j=-\infty}^{\infty} c_j \xi_{t-j}$

$\quad\quad\quad$ with $\{\xi_t\}$ a weak white noise and $\Sigma_{j=-\infty}^{\infty} c_j^2 < \infty$

Here $\{\xi_t\}$ is defined to be a weak white noise if

$\quad\quad E(\xi_t^2) = \phi^2 \quad$ and $\quad E(\xi_t \xi_s) = 0 \quad (t \neq s).$

A sufficient condition for the existence of a spectral density is $\Sigma_{j=1}^{\infty} |\rho_j| < \infty$ where ρ_j is the j-th lag autocorrelation. Also it is known that x_t has a one-sided moving average expression

(1.2) $\quad x_t = \Sigma_{j=0}^{\infty} c_j \xi_{t-j}$

$\quad\quad\quad$ with $\{\xi_t\}$ a weak white noise and $\Sigma_{j=0}^{\infty} c_j^2 < \infty$

if and only if $\{x_t\}$ has a spectral density $h(\omega)$ and it satisfies

(1.3) $\quad \int_{-\pi}^{\pi} |\log f(\omega)| \, d\omega < \infty$

(2) We assume that $\{x_t\}$ has an expression of (1.2). But the expression (1.2) is not unique. Here we construct a special expression. Let $\Psi_{t-1} = \{x_s : s = t-1, \ t-2, \cdots\}$ and $\Psi_{t-1, p} = \{x_s : s = t-1, \ t-2, \cdots, t-p\}$, and define

(1.4) $\quad \varepsilon_s = x_s - \hat{x}_s \quad$ with $\quad \hat{x}_s = E[x_s | \Psi_{s-1}]$

$(s = t, \ t-1, \cdots)$. Then

(1.5) $\quad \sigma^2 \equiv E(\varepsilon_s^2) \quad$ for all s and $\quad E(\varepsilon_s \hat{x}_s) = 0$

(see Anderson (1971) p.420), and letting $b_j = E(x_t \varepsilon_{t-j})/\sigma^2$, we obtain

(1.6) $x_t = \Sigma_{j=0}^{\infty} b_j \varepsilon_{t-j}$

with $\{\varepsilon_t\}$ a weak white noise and $\Sigma_{j=0}^{\infty} b_j^2 < \infty$,

which is nothing but an expression of (1.2). On the other hand x_t is orthogonally decomposed as

(1.7) $x_t = \hat{x}_{t,p} + \varepsilon_{t,p}$

with $\hat{x}_{t,p} = E[x_t | \Psi_{t-1,p}]$ and $E(\hat{x}_{t,p} \varepsilon_{t,p}) = 0$

and it holds that

(1.8) $\lim_{p \to \infty} E(\hat{x}_{t,p} - \hat{x}_t)^2 = \lim_{p \to \infty} E(\varepsilon_{t,p} - \varepsilon_t)^2 = 0.$

Also since

(1.9) $E(x_t^2) \geq E(\hat{x}_t^2) \geq E(\hat{x}_{t,p}^2)$ and

$E(\hat{x}_{t,p} - \hat{x}_t)^2 \geq E(\hat{x}_t^2) - E(\hat{x}_{t,p}^2) \geq 0,$

(1.10) $\lim E(\hat{x}_{t,p}^2) = E(\hat{x}_t^2)$

From the definition of $\Psi_{t-1,p}$ and from (1.7), $\hat{x}_{t,p} = \Sigma_{i=1}^{P} d_{ip}$ x_{t-i} for some d_{ip}'s and hence from (1.9)

(1.11) $\delta_p \equiv E(\hat{x}_{t,p}^2) = \Sigma_{i=1}^{P} \Sigma_{j=1}^{P} d_{ip} d_{jp} \gamma_{i-j}$

$\leq E(\hat{x}_t^2) \leq E(x_t^2) = \gamma_0,$

where $\gamma_j = \text{Cov}(x_t, x_{t-j})$. Therefore from (1.10)

(1.12) $\lim_{p \to \infty} \delta_p = E(\hat{x}_t^2) \equiv \delta < \infty$ (independent of t).

(3) Now suppose that $\{x_t\}$ is a normal stationary process of the expression (1.6). Then the characteristic function of $\hat{x}_{t,p}$ is

$\phi_p(z) = E[\exp(i z \hat{x}_{t,p})] = \exp(-\frac{z^2}{2} \delta_p)$

and let $\phi(z)$ be the characteristic function of $\hat{x}_t$. By (1.8), $\hat{x}_{t,p} \to \hat{x}_t$ in distribution and hence

$$\lim_{p\to\infty} \psi_p(z) = \exp(-\frac{z^2}{2}\delta) = \phi(z),$$

which implies that $\hat{x}_t$ is normal with mean 0 and variance δ.

Similarly to show that $(\hat{x}_t, \hat{x}_{t-1}, \cdots, \hat{x}_{t-k})$ is normal for any k, let

$$\hat{y}_t = \Sigma_{i=0}^k c_i \hat{x}_{t-i} \quad \text{and} \quad \hat{y}_{t,p} = \Sigma_{i=0}^k c_i \hat{x}_{t,p},$$

where c_i's are arbitrary. Then it is easy to see

$$E(\hat{y}_{t,p} - \hat{y}_t)^2 \leq \Sigma_{i=0}^k c_i^2 E(\hat{x}_{t,p} - \hat{x}_t)^2 \to 0 \quad \text{as} \quad p\to\infty.$$

Hence in the same way as above, $\hat{y}_t$ is shown to be normal for any $(c_0, c_1, \cdots, c_k)$, implying that $(\hat{x}_t, \hat{x}_{t-1}, \cdots, \hat{x}_{t-k})$ is jointly normal, which in turn implies that $(\varepsilon_t, \varepsilon_{t-1}, \cdots, \varepsilon_{t-k})$ is jointly normal as $\varepsilon_t = x_t - \hat{x}_t$. Since $E(\varepsilon_r \varepsilon_s) = 0$ $(r \neq s)$, $\{\varepsilon_t\}$ iid normal. Therefore a normal stationary process of one-sided moving average expression (1.6) is a linear process.

Chapter 4

1 Kalman filter theory

I. Let us review some results on normal distribution. Suppose

$$(1.1) \quad u \equiv \begin{pmatrix} u_1 \\ u_2 \\ u_3 \end{pmatrix} \sim N(\begin{pmatrix} 0 \\ 0 \\ 0 \end{pmatrix}, \begin{pmatrix} \Sigma_{11} & \Sigma_{12} & \Sigma_{13} \\ \Sigma_{21} & \Sigma_{22} & \Sigma_{23} \\ \Sigma_{31} & \Sigma_{32} & \Sigma_{33} \end{pmatrix}).$$

Lemma 1.1. (1) u_2 given $u_1 \sim N(\Sigma_{21}\Sigma_{11}^{-1}u_1, \Sigma_{22\cdot1})$ where $\Sigma_{22\cdot1} = \Sigma_{22} - \Sigma_{21}\Sigma_{11}^{-1}\Sigma_{21}$.

(2) Let $v = u_2 - \Sigma_{21}\Sigma_{11}^{-1}u_1$ and $\Delta = \Sigma_{32} - \Sigma_{31}\Sigma_{11}^{-1}\Sigma_{12}$. Then v and u_1 are independent and

$$\begin{pmatrix} v \\ u_3 \end{pmatrix} \sim N\left(\begin{pmatrix} 0 \\ 0 \end{pmatrix}, \begin{pmatrix} \Sigma_{22\cdot1} & \Delta' \\ \Delta & \Sigma_{33} \end{pmatrix} \right) \quad \text{with}$$

(3) $E[u_3|u_1, u_2] = E[u_3|u_1] + E[u_3|v]$ with

$$E[u_3|u_1] = \Sigma_{31}\Sigma_{11}^{-1}u_1 \quad \text{and} \quad E[u_3|v] = \Delta\Sigma_{22\cdot1}^{-1}v.$$

Proof.   (1) and (2) are clear.   Note $\mathrm{Cov}(v, u_1) = 0$.   For (3), use

$$\begin{pmatrix} \Sigma_{11} & \Sigma_{12} \\ \Sigma_{21} & \Sigma_{22} \end{pmatrix}^{-1} = \begin{pmatrix} I & -\Sigma_{12}^{-1}\Sigma_{12} \\ 0 & I \end{pmatrix} \begin{pmatrix} \Sigma_{11}^{-1} & 0 \\ 0 & \Sigma_{22\cdot1}^{-1} \end{pmatrix} \begin{pmatrix} I & 0 \\ -\Sigma_{21}\Sigma_{11}^{-1} & I \end{pmatrix}$$

in $E[u_3|u_1, u_2] = (\Sigma_{31} \quad \Sigma_{32}) \begin{pmatrix} \Sigma_{11} & \Sigma_{12} \\ \Sigma_{21} & \Sigma_{22} \end{pmatrix}^{-1} \begin{pmatrix} u_1 \\ u_2 \end{pmatrix}.$

II. Now our model is given by

(1.2) $\begin{aligned} f_t &= B_t f_{t-1} + \eta_t \\ x_t &= A_t f_t + \varepsilon_t \end{aligned}$, $\begin{pmatrix} \eta_t \\ \varepsilon_t \end{pmatrix}$ iid $N\left(0, \begin{pmatrix} \Omega & 0 \\ 0 & \Phi \end{pmatrix}\right)$,

where all A_t's, B_t's, Ω and Φ are known.   Assume for the initial value f_0, $f_0 \sim N(\mu_0, \Omega_0)$.   Define the conditional MSE (mean square error) matrix of a filter $\widetilde{f}_{t|t}$ given $X_t = (x_t', x_{t-1}', \cdots, x_0')$: $p(t+1)$ by $P_{t|t}(\widetilde{f}_{t|t})$, where

(1.3) $P_{s|t}(\widetilde{f}) = E[(f_s - \widetilde{f})(f_s - \widetilde{f})' | X_t],$

where $\widetilde{f}_{t|t}$ is a filter of f_t based on the past and present obser- vation vector X_t.   Then it is easy to see that the best filter is given by

(1.4) $\hat{f}_{t|t} = E[f_t|X_t].$

Similarly the best prediction of f_{t+1} given X_t is given by

(1.5) $\hat{f}_{t+1|t} = E[f_{t+1}|X_t] = B_t \hat{f}_{t|t}$

under the MSE matrix $P_{t+1|t}(\widetilde{f}_{t+1|t})$. Let

(1.6) $\quad x_{t|t-1} = E[x_t | X_{t-1}]$ and $\quad \nu_t = x_t - \hat{x}_{t|t-1}.$

Then ν_t is the new information at t which is not contained in the past data X_{t-1}, and it is called an innovation at t. By Lemma 1. 1(2), ν_t and X_{t-1} are independent since x_t and X_{t-1} are jointly normal. Also from (1.3)

(1.7) $\quad \hat{x}_{t|t-1} = A_t \hat{f}_{t|t-1} = E[A_t f_t + \varepsilon_t | X_{t-1}]$
$\qquad\qquad = E[A_t B_t f_{t-1} + A_t \eta_t + \varepsilon_t | X_{t-1}]$
$\qquad\qquad = A_t B_t \hat{f}_{t-1|t-1} = A_t \hat{f}_{t|t-1}.$

Hence by Lemma 1.1(3), (1.6) and (1.7), the best filter in (1.4) is

(1.8) $\quad \hat{f}_{t|t-1} = \hat{f}_{t|t-1} + K_t \nu_t$

because

$\qquad \hat{f}_{t|t} = E[f_t | X_{t-1}, \; x_t] = E[f_t | X_{t-1}] + E[f_t | \nu_t]$
$\qquad\quad = \hat{f}_{t|t-1} + K_t \nu_t = \hat{f}_{t|t-1} + K_t[x_t - A_t B_t \hat{f}_{t-1|t-1}],$

where the joint normality of $(x_t, \; X_{t-1}, \; f_t)$ was used. Here K_t is a constant which depends on the covariance matrix of $(f_t, \; \nu_t, \; X_{t-1})$. The equation (1.8) gives a recurrsive formula for the best filter $\hat{f}_{t|t}$ given $\hat{f}_{t|t-1}$ and x_t.

III. Let the residual due to the estimation of f_t by $f_{t|t}$ be

(1.9) $\quad e_{t|t} = f_t - \hat{f}_{t|t},$

which gives the MSE matrix $P_{t|t} \equiv \hat{P}(f_{t|t}) = E[e_{t|t} e_{t|t}']$, and let the error due to the prediction of f_t by $\hat{f}_{t|t-1}$ be

(1.10) $\quad e_{t|t-1} = f_t - \hat{f}_{t|t-1},$

which gives the prediction error matrix $P_{t|t-1} \equiv P(\hat{f}_{t|t-1}) =$

$E[\,e_{t|t-1}\,e_{t|t-1}{}'\,]$.   Then by (1.8)

(1.11) $e_{t|t} = e_{t|t-1} - K_t \nu_t = e_{t|t-1} - E[\,e_{t|t-1}|\nu_t]$

since $E[\,e_{t|t-1}|X_t] = E[\,e_{t|t-1}|\nu_t] = K_t \nu_t$.   Here note that by (1.10), $E[\,e_{t|t-1}|X_{t-1}] = 0$ and that ν_t and $e_{t|t-1}$ are jointly normal.   Therefore by Lemma 1.1(2) and (1.11), $e_{t|t}$ and ν_t are independent with $E[\,e_{t|t}] = 0$, implying

(1.12) $\begin{aligned}
0 &= E[\,e_{t|t}\,\nu_t{}'] = E[(\,e_{t|t-1} - K_t \nu_t)\nu_t{}'] \\
&= E[\,e_{t|t-1}(A_t\,e_{t|t-1} + \varepsilon_t)'] - K_t E(\nu_t \nu_t{}') \\
&= P_{t|t-1} A_t{}' - K_t[A_t P_{t|t-1} A_t{}' + \varPhi].
\end{aligned}$

In fact, by (1.3) and (1.7)

(1.13) $\begin{aligned}
\nu_t &= x_t - \hat{x}_{t|t-1} = A_t(\,f_t - \hat{f}_{t|t-1}) + \varepsilon_t \\
&= A_t\,e_{t|t-1} + \varepsilon_t \quad \text{and}
\end{aligned}$

$\begin{aligned}
E(\nu_t \nu_t{}') &= E[(A_t\,e_{t|t-1} + \varepsilon_t)(A_t\,e_{t|t-1} + \varepsilon_t)'] \\
&= A_t P_{t|t-1} A_t{}' + \varPhi.
\end{aligned}$

Consequently

(1.14) $K_t = P_{t|t-1} A_t{}'(A_t P_{t|t-1} A_t{}' + \varPhi)^{-1}.$

Of course (1.12) and (1.8) give the best filter.

Ⅳ. We evaluate $P_{t|t} = E[\,e_{t|t}\,e_{t|t}{}']$.   Using (1.8), (1.12) an (1.13),

$\begin{aligned}
P_{t|t} &= E[(\,e_{t|t-1} - K_t \nu_t)(\,e_{t|t-1} - K_t \nu_t)'] \\
&= P_{t|t-1} - K_t E(\nu_t\,e_{t|t-1}) - E(\,e_{t|t-1}\nu_t{}')K_t{}' \\
&\qquad\qquad\qquad\qquad + K_t E(\nu_t \nu_t{}')K_t \\
&= (I - K_t A_t)P_{t|t-1}
\end{aligned}$

with K_t in (1.14).   This gives a recurrsive formula for $P_{t|t}$ and $P_{t|t-1}$.

V. Finally the best predictor which minimizes the MSE matrix $P_{t+1|t}(\tilde{f}_{t+1|t})$ is given by $\hat{f}_{t+1|t} = E[f_{t+1}|X_t]$. From $f_{t+1} = B_t f_t + \eta_t$, $\hat{f}_{t+1|t} = B_{t+1}\hat{f}_{t|t}$. Hence the prediction MSE matrix $P_{t+1|t}$ is

$$P_{t+1|t} = E[(B_{t+1} e_{t|t} + \eta_t)(B_{t+1} e_{t|t} + \eta_t)']$$
$$= B_{t+1} P_{t|t} B_{t+1}' + \Omega.$$

This proves the results on the Kalman filter procedure.

Chapter 5

1 Proof of Theorem 2.1

Assume that the spectral density matrix $H(\omega)$ of $\{x_t\}$ exists;

$$(1.1)\quad H(\omega) = \frac{1}{2\pi}\sum_{k=-\infty}^{\infty}\exp(i\omega k)\Sigma(k).$$

Here when $p>1$, $H(\omega)$ is not symmetric but Hermitian, i.e., $H(\omega)' = \overline{H(\omega)}$ because of $\Sigma(-k) = \Sigma(k)'$.

Now first assume that $\{x_t\}$ is time-reversible. Then $\Sigma(k)' = \Sigma(k)$ and hence $H(\omega)$ is symmetric and real. Hence there exists a $p\times p$ orthogonal matrix $A(\omega)$ which diagonalizes $H(\omega)$ as

$$(1.2)\quad A(\omega)' H(\omega) A(\omega) = \text{diag}\{h_1(\omega), \cdots, h_p(\omega)\} \equiv D_k(\omega),$$

where $\gamma_1(0) \geqq \gamma_2(0) \geqq \cdots \geqq \gamma_p(0)$ with

$$\gamma_j(k) = \int_{-\pi}^{\pi}\exp(-k i\omega)h_j(\omega)d\omega.$$

For the identifiability of $h_j(\omega)$'s, we need the strict inequalities;

$$(1.3)\quad \gamma_1(0) > \cdots > \gamma_p(0).$$

In fact, if (1.3) holds, $A(\omega)$ is unique up to transformation $A(\omega)\rightarrow A(\omega)E$ with $E\in\varepsilon$, where

(1.4) $\varepsilon = \{E=\mathrm{diag}\{e_1,\cdots,e_p\}: e_i=1\,\mathrm{or}\,-1\}$.

Further assume that $A(\omega)\equiv A$ does not depend on ω. This assumption holds if and only if $\Sigma(k)$'s are simultaneously diagonalized. Then by (1.2)

(1.5) $H(\omega)=A\,D_k(\omega)\,A' = h_1(\omega)\,\alpha_1\alpha_1'+\cdots+h_p(\omega)\,\alpha_p\alpha_p'$,

where $A=[\alpha_1,\cdots,\alpha_p]$. Under (1.5), let $\{f_{jt}\}$ represent a process with spectral density $h_j(\omega)$. Then we obtain the MTV model as a model representing the spectral density matrix $H(\omega)$; $x_t=\mu+\alpha_1 f_{1t}+\cdots+\alpha_p f_{pt}$ with the conditions
  (A) $F_j=\{f_{jt}\}$ is stationary ($j=1,\cdots,p$),
  (B) F_j and F_k are uncorrelated ($j\neq k$),
  (C) $\gamma_1(0)>\cdots>\gamma_p(0)$ when (1.5) is assumed and
  (D) $A'A=I$ where A is unique up to $\{AE\}$ with $E\in\varepsilon$.
    This proves Theorem 2.1.

Chapter 7

1 Proof of (4.3).   Rewrite (4.1) in Chapter 7 as

(1.1) $x_t=\alpha_0+\alpha_1 f_{1t}+\cdots+\alpha_q f_{qt}+\varepsilon_t$.

Let $A^*=[1,\alpha_1,\cdots,\alpha_q]: N\times(q+1)$ and $M=A^*(A^{*\prime}A^*)^{-1}A^{*\prime}$ and decompose α_0 as

(1.2) $\alpha_0=d+e$ with $d=(I-M)\alpha_0$ and $e=M\alpha_0$.

Then it is easy to see

(1.3) $d'1=0,\quad d'\alpha_j=0\quad(j=1,\cdots,q)$.

Here regarding $\theta\,d$ as a portfolio with θ constant, from (1.1) and (1.3)

$$r_t(\boldsymbol{d})=\boldsymbol{d}'\,\boldsymbol{x}_t=\theta\{\boldsymbol{d}'\,\boldsymbol{\alpha}_0+\boldsymbol{d}'\,\boldsymbol{\varepsilon}_t\}=\theta\{\boldsymbol{d}'\,\boldsymbol{d}+\boldsymbol{d}'\,\boldsymbol{\varepsilon}_t\}$$

and hence the mean and variance of the portfolio are

$$\mu_t(\boldsymbol{d})=E(r_t(\boldsymbol{d}))=\theta\,\boldsymbol{d}'\,\boldsymbol{d}, \text{ and}$$
$$\sigma_t^2(\boldsymbol{d})=\mathrm{Var}(r_t(\boldsymbol{d}))=\theta^2\boldsymbol{d}'\,\boldsymbol{\Phi}\boldsymbol{d}\leqq K\theta^2\boldsymbol{d}'\,\boldsymbol{d},$$

respectively where $K=\max|\phi_{ii}|$. Take $\theta=(\boldsymbol{d}'\,\boldsymbol{d})^{-2/3}$ to get

$$\mu_t(\boldsymbol{d})=(\boldsymbol{d}'\,\boldsymbol{d})^{1/3} \text{ and } \sigma_t^2(\boldsymbol{d})\leqq K(\boldsymbol{d}'\,\boldsymbol{d})^{-1/3}.$$

Therefore if $\boldsymbol{d}'\,\boldsymbol{d}\to\infty$ as $N\to\infty$, then $\mu_t(\boldsymbol{d})\to\infty$ and $\sigma_t^2(\boldsymbol{d})\to0$, which will imply an arbitrage opportunity because $\boldsymbol{d}'\,1=\Sigma_{i=1}^N d_i=0$ means zero (no cost) portfolio (short=long). Thus by assumption we have $\boldsymbol{d}'\,\boldsymbol{d}<\infty$ as $N\to\infty$. Further since $\boldsymbol{d}=\boldsymbol{\alpha}_0-\boldsymbol{e}$ and $\boldsymbol{e}=\lambda_0 1+\lambda_1\boldsymbol{\alpha}_1+\cdots+\lambda_q\boldsymbol{\alpha}_q$ from its definition (note $\boldsymbol{\lambda}=(\lambda_0, \lambda_1,\cdots,\lambda_q)'=(\boldsymbol{A}^{*'}\boldsymbol{A}^*)^{-1}\boldsymbol{A}^{*'}\boldsymbol{\alpha}_0)$, $\boldsymbol{d}=\boldsymbol{\alpha}_0-\lambda_0 1-\Sigma_{j=1}^q\lambda_j\boldsymbol{\alpha}_{.j}$. Hence

$$\Sigma_{i=1}^N(\alpha_{i0}-\lambda_0-\Sigma_{j=1}^q\lambda_j\alpha_{ij})^2<\infty \text{ as } N\to\infty,$$

which in turn implies $\alpha_{i0}\doteqdot\lambda_0+\Sigma_{j=1}^q\lambda_j\alpha_{ij}$ for i large. But by symmetry it holds for all i.

Chapter 12

1 Geometric Brownian motion

A stochastic process $\{Z_t\}\equiv\{Z_t: t\in[0, \infty)\}$ with continuous time is said to be a Brownian motion with drift μ and variance σ^2, denoted by $\mathrm{BM}(\mu, \sigma^2)$, if

(i) $Z_0=0$

(ii) for any $0\leqq t_0\leqq t_1<\cdots<t_N$, $U_{tk}\equiv Z_{tk}-Z_{tk-1}$'s $(k=1,\cdots, N)$ are independently normally distributed with mean $\mu\Delta t_k$ and variance $\sigma^2\Delta t_k$ where $\Delta t_k=t_k-t_{k-1}$.

In particular, if $\{Z_t\}\sim$BM$(0,1)$, $\{Z_t\}$ is called a Wiener process. Assume $\{Z_t\}\sim$BM(μ, σ^2). Then

$$Z_{t\,k} = U_{t\,k} + U_{t\,k-1} + \cdots + U_{t0} \quad \text{with} \quad U_{t0} = Z_{t0}.$$

Hence if $\Delta t_k \equiv \Delta t$ ($k=1,\cdots, N$) and $t_0 = \Delta t$, $Z_{t\,k}$ is a sum of iid normal variates. Also in general Cov$(Z_{t\,i}, Z_{t\,j}) = \sigma^2 \min(t_i, t_j)$.

Lemma 1.1. $\{Z_t\}\sim$BM$(0,1)$ if and only if (i) (ii) and (iii) hold:
(i) $Z_{t+s} - Z_t$ is independent of $\{X_0: \tau \le t\}$ ($s>0$).
(ii) The distribution of $Z_{t+s} - Z_t$ is independent of t.
(iii) $\lim_{s\downarrow 0} \frac{1}{s} P(|Z_{t+s} - Z_t| > \delta) = 0$ for any $\delta > 0$.

Proof.   See Breiman (1968).

Ito Calculus
   Assume $\{Z_t\}\sim$BM$(0,1)$. Define the notation

$$(1.1) \quad d S_t = \alpha(t, S_t) d t + \beta(t, S_t) d Z_t$$

by the following stochastic integral

$$(1.2) \quad S_t - S_a = \int_a^t \alpha(s, S_s) d s + \int_a^t \beta(s, S_s) d Z_s,$$

where the first term of the right side is regarded as a Rieman integral with $\alpha(s, S_s)$ continuous in s. The second term of the right side is the stochastic integral defined by Ito in terms of in-probability concept. However, recently it is defined in terms of L^2-concept with martingale process (see Chung and William (1983)). Here the idea of Ito integral is that $\gamma(s) = \beta(s, S_s)$ is approximated by a simple function $\gamma_n(s)$ and the integral in the second term is defined to be a in-probability limit of

$$\int_a^t \gamma_n(s) d Z_s = \sum_{i=1}^n \gamma_n(t_{i-1})[Z_{t\,i} - Z_{t\,i-1}]$$

as $\Delta t_i \to 0$, where $a = t_0 < t_1 < \cdots < t_n = t$. In particular.

$$\int_0^t d Z_t = Z_t - Z_0 = Z_t.$$

Ito's Lemma   Let $u(t, x): [0, T] \times R \to R$ be a continuously twice differentiable function with partial derivatives u_t, u_x, u_{xx}, u_{xt}, and u_{tt}. Then when $\{S_t\}$ follows the process (1.1) via (1.2), then $Y_t = u(t, S_t)$ follows

(1.3) $d Y_t = \alpha'(t, S_t) d t + \beta'(t, S_t) d Z_t$, where

(1.4) $\alpha'(t, S_t) = u_t(t, S_t) + u_x(t, S_t) \alpha(t, S_t)$
$\qquad\qquad + \frac{1}{2} u_{xx}(t, S_t) \beta(t, S_t)^2$ and
$\qquad \beta'(t, S_t) = u_x(t, S_t) \beta(t, S_t).$

As computational formulae in Taylor expansion, one can use

$$d t \times d t = 0, \quad d t \times d Z_t = 0.$$
$$d Z_t \times d Z_s = 0 \ (t \ne s) \text{ and } (d Z_t)^2 = d t.$$

(1) Suppose $\{S_t\}$ follows the process

(1.5) $d S_t = \mu S_t d t + \sigma S_t d Z_t.$

Then apply Ito's lemma to $Y_t = \log S_t$ to get

$$d Y_t = (\mu - \frac{1}{2}\sigma^2) d t + \sigma d Z_t,$$

which is by definition equivalent to

$$Y_t - Y_0 = \int_0^t (\mu - \frac{1}{2}\sigma^2) d s + \sigma \int_0^t d Z_t.$$

Hence we obtain a form of geometric Brownian motion.

(1.6) $S_t = S_0 \exp[(\mu - \frac{1}{2}\sigma^2) t + \sigma Z_t].$

When $\{S_t\}$ follows (1.6), then

$$x_t = \log S_t - \log S_{t-\Delta t} = (\mu - \tfrac{1}{2}\sigma^2)\Delta t + \sigma(Z_t - Z_{t-\Delta t}).$$

Hence if $\Delta t = 1$, returns x_t's are iid $N(\mu - \tfrac{1}{2}\sigma^2, \sigma^2)$, implying that $\{\log S_t\}$ is a normal strong random walk (see Chapter 2).

(2) Assume $\{S_t\}$ follows (1.5), and let $Y_t = C(t, S_t)$. Then by Ito's Lemma we get $dY_t = \alpha_t' \, dt + \sigma_t' \, dZ_t$ with

$$\alpha_t' = C_t(t, S_t) + C_x(t, S_t)\mu S_t + \tfrac{1}{2}C_{xx}(t, S_t)(\sigma S_t)^2 \quad \text{and}$$
$$\sigma_t' = C_x(t, S_t)\sigma S_t.$$

2 Outline of the proof of Theorem 3.1

(1) Derivation of $\mathrm{Var}(C_T^*) = E(C_T^{*2}) - C_T^2$. From $C_T^* = e^{-r\tau}$ $\max(S_T - K, 0)$ and $S_T = S_T(z) = S_t\exp(\mu\tau + \sigma\sqrt{\tau}\,z)$ with $z \sim N(0,1)$, let $f(z)$ be the pdf of z and set $A = \{z \mid S_T(z) > K\}$. Then

$$E(C_T^{*2}) = e^{-2r\tau}\int_A (S_T(z) - K)^2 \, f(z) \, dz$$
$$= e^{-2r\tau}\left[\int_A S_T(z)^2 \, f(z) \, dz - 2K\int_A S_T(z) \, f(z) \, dz \right.$$
$$\left. + K^2 \int_A f(z) \, dz\right]$$
$$= \mathrm{I} - \mathrm{II} + \mathrm{III}.$$

Here, letting $y = 2\sigma\sqrt{\tau} - x$ and $d = [\log(S_t/K) + \mu\tau]/\sigma\sqrt{\tau}$, A is equivalent to $B = \{y \mid d + \sigma\sqrt{\tau} > y\}$. Hence $\mathrm{I} = S_t^2\exp(\sigma^2\tau)$ $\Phi(d + \sigma\sqrt{\tau})$, $\mathrm{II} = -2K\exp(-r\tau)\Phi(d)$, and $\mathrm{III} = K^2\exp(-2r\tau)$ $\Phi(d - \sigma\sqrt{\tau})$, where $r = \mu + \tfrac{1}{2}\sigma^2$ was used.

(2) Derivation of $\mathrm{Cov}(S_T^*, C_T^*) = E(S_T^* C_T^*) - C_t S_t$. Use

$$E(C_T^* S_T^*) = \int_A (S_T^*(z) - K) S_T^*(z) \, f(z) \, dz$$
$$= S_t^2\exp(\sigma^2\tau)\Phi(d + \sigma\sqrt{\tau}) - K\exp(-r\tau)S_t\Phi(d).$$

(3) (1), (2) and $\mathrm{Var}(S_T^*) = S_t^2[\exp(\sigma^2\tau) - 1]$ gives $\mathrm{Correl}(S_T^*, C_T^*)$.

REFERENCES

Akgiray, V. (1989). "Conditional heteroscedasticity in time series of stock returns: Evidence and forecasts", *Jour. Business* **62**, 55 -80.

Anderson, T.W. (1971). *The Statistical Analysis of Time Series*, John Wiley & Sons, New York.

Anderson, T.W. and A.M. Walker (1964). "On the asymptotic distribution of the autocorrelations of a sample from a linear stochastic process", *Ann. Math. Statist.* **35**, 1296-1303.

Baba, Y. (1990). "The ARCH model: its introduction and an analysis of time varying risk premium", *Jour. Japan Statist. Soc.* **20**, 217-226.

Ball, C.A. and W.N. Torous (1984). "The maximum likelihood estimation of security price volatility: Theory, evidence and application to option pricing", *Jour. Business* **57**, 97-112.

Bassett, G.W.Jr., V.G. France and S.R. Pliska (1990). "Kalman filter estimates of the MMI cash-futures spread on October 19 and 20, 1987", *Jour. Account. Finance.*

Bhaskara Rao M., T. Subba Rao and A.M. Walker (1983). "On the existence of some bilinear time series models", *Jour. Time Series Anal.* **4**, 95-110.

Bierwag, G.O. (1986). *Duration Analysis*, Ballinger Publishing Co.

Black, F. and M. Scholes (1973). "The pricing of options and corporate liabilities", *Jour. Political Economy* **81**, 637-659.

Bollerslev, T.P. (1986). "Generalized autoregressive conditional heteroscedasticity", *Jour. Econometrics* **31**, 307-327.

Bollerslev, T.P. (1988). "On the correlation structure for the generalized autoregressive conditional heteroscedasticity process", *Jour. Time Series Anal.* **9**, 121-132.

Bollerslev, T.P., R.F. Engle and J.M. Woodridge (1988). "A capital asset pricing model with time varying covariance", *Jour. Political Economy* **96**, 116-131.

Boothe, P. and D. Glassman (1987). "The statistical distribution of exchange rates: Empirical evidence and economic implications", *Jour. International Economics* **22**, 297-319.

Boyle, P.P. and A.L. Ananthanarayanan (1977). "The impact of variance estimation in option valuation models", *Jour. Financial Economics* **5**, 375-388.

Boyle, P.P. (1991).   "Multi-asset path-dependent options", unpublished.

Breiman, L. (1968).   *Probability*, Addison Wesley, New York.

Brown, R.L., J. Durbin and J.M. Evans (1975).   "Techniques for testing the constancy of regression relationships over time", *Jour. Roy. Statist. Soc.*, **37B**, 149-192.

Brown, S.J. and T. Otsuki (1988).   CAPMD: "A model of risk and return in the Japanese equity market", Unpublished.

Brown, S.J. and M.I. Weinstein (1983).   "A new approach to testing asset pricing models: The bilinear paradigm", *Jour. Business* **38**, 711-742.

Brown, S.J. and T. Otsuki (1990).   "Macroeconomic factors and the Japanese equity market: the CAPMD project", in *Japanese Capital Market* (ed. by Elton, E.J. and Bruber, M.J.).

Campbell, J.Y. and R.J. Shiller (1984).   "A simple account of the behavior of long-term interest rates", *Amer. Econ. Review* **74**, 44 -48.

Chen, N.F., R. Roll and S.A. Ross (1986).   "Economic forces and the stock market", *Jour. Business* **59**, 383-403.

Chiras, D.P. and S. Manaster (1978).   "The information content of option prices and a test of market efficiency", *Jour. Financial Economics* **6**, 213-234.

Chung, K.L. (1977).   *A Course in Probability*, Academic Press, New York.

Chung and William (1983).   *Introduction to Stochastic Integration*, Birkhauser, Boston.

Clark, P.K. (1973).   "A subordinated stochastic process model with finite variance for speculative prices", *Econometrica* **41**, 135-155.

Cox, J.C. and M. Rubinstein (1985).   *Options Markets*, Prentice-Hall, Englewood Cliffs, New Jersey.

Darroch, J.N. (1965). "An optimal property of principal components", *Ann. Math. Statist.* **36**, 1579-1582.

Dickey, D.A. and W.A. Fuller (1981).   "Likelihood ratio statistics for autoregressive time series unit root", *Econometrica* **49**, 1057-1072.

Domowitz, I. and C.S. Hakkio (1985). "Conditional variance and the risk premium in the foreign exchange market", *Jour. International Economics* **19**, 47-66.

Elton, E.J. and M.J. Gruber (1988). "A multi-risk model of the Japanese stock market", *Japan and the World Economy* **1**, 21-44.

Engle, R.F. (1982). "Autoregressive conditional heteroscedasticity with estimates of the variance of United Kingdom inflation", *Econometrica* **50**, 987-1007.

Engle, R.F. and D.F. Kraft (1983). "Multiperiod forecast error variances of inflation estimated from ARCH models", in *Applied Time Series Analysis of Economic Data*, ed. A. Zellner, Washington, D.C.:U.S. Bureau of the Census.

Engle, R.F., D.M. Lilien and R.P. Robins (1987). "Estimating time varying risk premia in the term structure: The ARCH-M model", *Econometrica* **55**, 391-407.

Fama, E.F. (1965). "The behavior of stock-market prices", *Jour. Business* **38**, 34-105.

Fielitz, B.D. and J.P. Rozelle (1983). "Stable distributions and the mixtures of distributions hypotheses for common stock returns", *Jour. Amer. Statist. Assoc.* **78**, 28-36.

Friedman, D. and S. Vandersteel (1982). "Short-run fluctuations in foreign exchange rates: Evidence from the data 1973-79", *Jour. International Economics* **13**, 171-186.

Gallent, A.R. (1975). "Seeming unrelated nonlinear regressions", *Jour. Econometrics* **3**, 35-50.

Garman, M.B. and M.J. Klass (1980). "On the estimation of security price volatilities from historical data", *Jour. Business* **53**, 67-78.

Geweke, J.F. (1977). "The dynamic factor analysis of economic time series models", *Latent Variables in Socio-economic Models*, North-Holland, 365-383.

Geweke, J.F. and K.J. Singleton (1981). "Maximum likelihood ' confirmatory' factor analysis of economic time series", *International Econ. Review* **22**, 37-54.

Geweke, J. and S. Porter-Hudak (1983). "The estimation and application of long-memory time series models", *Jour. Time Series Anal.* **4**, 221-238.

Giddy, Ian H. and Gunter Dufey (1975). "The random behavior of flexible exchange rates: Implications for forecasting", *Jour.*

International Business Studies **6**, 1–30.

Granger, C.W.J. and A.P.Anderson (1978). *An Introduction to Bilinear Time Series Models*, Vandenboek and Ruprecht, Gottingen.

Granger, C.W.J. and P. Newbold (1976). "Forecasting transformed series", *Jour. Royal Statist. Soc.* **38B**, 189–203.

Granger, C.W.J. and R. Joyeux (1980). "An introduction to long-memory time series models and fractional differencing", *Jour. Time Series Anal.* **1**, 15–29.

Greene, M.T. and B.D. Fielitz (1977). "Long-term dependence in common stock returns", *Jour. Financial Economics* **4**, 339–349.

Hagerman, R.L. (1978). "More evidence on the distribution of security returns", *Jour. Finance* **33**, 1213–1221.

Haggan, V. and T. Ozaki (1981). "Modelling nonlinear vibrations using an amplitude-dependent autoregressive time series model", *Biometrika* **68**, 189–196.

Hannan, H.J. (1970). *Multiple Time Series*, John Wiley & Sons, New York.

Harrison, J.M. and S.R. Pliska (1981). "Martingales and stochastic integrals in the theory of continuous trading", *Stochastic Process Appl.* **11**(3), 215–260.

Heath, D. (1989). "Contingent claim valuation with a random evolution of interest rates", Cornell University.

Heath, D. (1990). "Bond pricing and the term structure of interest rates: A discrete time approach", *Jour. Financial Quant. Anal.* **25**, 419–440.

Heath, D., R. Jarrow and Morton (1987). "Bond pricing and the term structure of interest rates: A continuous time approach". Cornell University.

Hinich, M.J. (1982). "Testing for Gaussianity and linearity of a stationary time series", *Jour. Time Series Anal.* **3**, 169–176.

Hosking, J.R.M. (1981). "Fractional differencing", *Biometrika*, **68**, 165–176.

Hurst, H.E. (1951). "Long-term storage capacity of reservoirs", *Trans. Amer. Soc. Civil. Engin.* **116**, 770–808.

Islam, S. (1982). "Statistical distribution of short-term exchange rate variation", *Research Paper* **8215**, FRB, New York.

Jarrow, R. and A. Rudd (1983). "Tests of an approximate option-valuation formula", in *Option Pricing*, ed. Menachem Brenner, 81-100, Lexington.

Jarrow, R. and A. Rudd (1983). *Option Pricing*, Irwin, Homewood.

Jones, R.C. (1987). *Stock Selection*, Goldman Sachs.

Kariya, T. (1987). "MTV model and its application to prediction of stock prices", *Proc. Second International Tampere Conference in Statistics* (ed. by Pukkila, T. and Puntanen, S.), 161-176.

Kariya, T. (1992). "Characterization of MTV model and its diagnostic checking". To appear from the Proceeding of the third Pacific Area Statistical Conference.

Kariya, T. and Y. Toyooka (1985). "Nonlinear version of the Gauss-Markov theorem and GLSE", *Multivar. Analysis* VI, 345-354.

Kariya, T. and B.K. Sinha (1989). *Robustness of Statistical Tests*, Academic Press, Boston.

Kariya, T. and Y. Matsue (1989). "Testing the random walk hypothesis of yen-dollar exchange rates in S.Taylor's model", *Discussion Paper Series* A No.198, Hitotsubashi University.

Kariya, T., Y. Tsukuda and J. Maru (1990). "Testing the random walk hypothesis for Japanese stock prices", *Tech. Rep.* 94, Graduate School of Business, University of Chicago.

Kariya, T. and Y. Tsukuda (1990). "Temporal aggregation of financial time series in Taylor model", Unpublished.

Kariya, T. and Katsuura, M. (1992). Testing the long term dependency for Japanese stock prices and Yen/Dollar exchange rates (in Japanese).

Kelker, D. (1970). "Distribution theory of spherical distribution and a location scale parameter generation", *Sankhya* A43, 419-430.

Kendall, M. and A. Stuart (1960). *The Theory of Statistics*, 2, 4th. ed. Charles Griffin, London.

King, B.F. (1966). "Market and industry factors in stock price behaviour", *Jour. Business* 28, 139-190.

Knif, J. (1989). *Parameter Variability in the Single Facdtor Market Model*, Finish Society of Sciences and Letters 40.

Kuwana, Y. and T. Kariya (1991). "LBI tests for multivariate normality in exponential power distributions", *Jour. Multivar. Anal.* 39, 117-134.

Latane, H.A. and R.J. Rendleman Jr. (1976). "Standard deviations of stock price ratios implied in option prices", *Jour. Finance* **31**, 3, 369-381.

Leland, H.E. (1980). "Who should buy portfolio insurance?", *Jour. Finance* **35**, 2, 581-598.

Litterman, R. and J. Scheinkman (1988). "Common factors affecting bond returns.", Goldman & Sachs.

Lo, A.W. (1991). "Long-term memory in stock market prices", *Econometrica*, **59**, 1279-1313.

Luukkonen, R., P. Saikkonen and T. Tcrasvirta (1988). "Testing linearity against smooth transition autoregressive models", *Biometrika* **75**, 491-499.

Mandelbrot, B. (1960). "The Pareto-Levvy law and the distribution of income", *International Econ. Review* **1**, 79-106.

Mandelbrot, B. (1963). "The variation of certain specurative prices", *Jour. Business* **36**, 394-419.

Mandelbrot, B. and J.W. Van Ness (1968). Fractional Brownian motions, fractional noise and applications, *SIAM Rev.* **10**, 422-437.

Markowitz, H.M. (1956). *Portfolio Selections*, Cowles Foundation.

McCulloch, J.H. (1971). "Measuring the term structure of interest rates", *Jour. Business* **44**, 19-31.

McElroy, M.B., E. Burmeister and K.D. Wall (1985). "Two estimators for the APT model when factors are measured", *Economic Letters* **19**, 271-275.

McElroy, M.B. and Burmeister (1988). "Arbitrage pricing theory as a restricted nonlinear multivariate regression model", *Jour. Business & Econ. Statist.* **6**, 29-42.

Milhoj, A. (1985). "The moment structure of ARCH process", *Scand. Jour. Statist.* **12**, 281-292.

Milhoj, A. (1987). "A conditional variance model for daily deviations of an exchange rate", *Jour. Business & Econ. Statist.* **5**, 99-103.

Nelson, D. (1989). "Conditional heteroscedasticity in asset returns", University of Chicago.

Nimmo-Smith, I. (1979). "Linear regressions and shpericity", *Biometrika* **66**, 390-392.

Okamoto, M.B. (1992). "Long range dependence of foreign exchange rate and estimation of parameter in fractionally difference process", *Economics Annually* **13**, 1-11, Hiroshima University.

Okamoto, M. and M. Kanazawa (1968). "Minimization of eigenvalues of a matrix and optimality of principal components", *Ann. Math.*

Statist. **39**, 859-863.

Ozaki, T. (1981). "Nonlinear threshold autoregressive models for nonlinear random vibrations", *Jour. Appl. Prob.* **18**, 443-451.

Pagano, A. (1977). "Estimation of models of autoregressive signal plus noise", *Ann. of Statistics* 2, 99-108.

Parkinson, M. (1980). "The extreme value method for estimating the variance of the rate of return", *Jour. Business* **53**, 61-65.

Patell, M. and M.A. Wolfson (1981). "The Ex Ante and Ex Post effects of quarterly earnings announcements reflected in option and stock prices", *Jour. Account. Research* 1, 434-458.

Peña, D. and G.E.P. Box (1987). "Identifying a simple structure in time series", *Jour. Amer. Statist. Assoc.* **82**, 836-843.

Perry, P.R. (1982). "The time variance relationship of security returns: Implications for the return-generating stochastic process", *Jour. Finance* 37,   857-870.

Petruccalli, J. and S.W. Woolford (1984). "A threshold AR(1) model", *Jour. Appl. Prob.* **21**, 270-286.

Petruccelli, J. and N. Davis (1986). "A portmanteau test for self-exciting threshold autoregressive-type nonlinearity in time series", *Biometrika* **73**, 687-694.

Pham, D.T. and L.T. Tran (1981). "On first-order bilinear time series models", *Jour. Appl. Prob.* **18**, 617-27.

Phillips, S.M. and C.W. Smith Jr. (1980). "Trading costs for listed options. The implications for market efficiency", *Jour. Financial Economics* **8**, 179-201.

Praetz, P.D. (1979). "Testing for a flat-spectrum on efficient market price data", *Jour. Finance* **34**, 645-658.

Press, S.J. (1973). *Applied Multivariate Analysis*, Holt, Rinehart and Winston.

Rao, C.R. and R. Boudreau (1985). "Prediction of future observations in a factor analytic type growth model", *Multivariate Analysis* VI (P.R. Krishnaiah, ed.), North-Holland, 449-466.

Roll, R. and S. Ross (1980). "An empirical investigation of the arbitrage pricing theory", *Jour. Finance* **35**, 1073-1102.

Rosenberg, B. (1973a). "A survey of stochastic parameter regression", *Ann. Econ. Social Measurement* 2, 381-397.

Rosenberg, B. (1973b). "Linear regression with randomly dispersed parameters", *Biometrika* **60**, 65-72.

Rosenberg, B. (1973c). "The analysis of a cross-section of time series by stochastically convergent parameter regression", *Ann.*

Econ. Social Measurement 2.

Rosenberg, B. (1973d). "The behavior of random variables with non-stationary variance and the distribution of securities prices", Unpublished.

Rosenberg, B. (1973e). "Random coefficient cross-section and time series models", *Ann. Econ. Social Measurement* 2.

Rosenberg, B. and W. McKibben (1973). "The prediction of systematic and specific risk in common stocks", *Jour. Financial Quant. Anal.* 317-333.

Ross, S.A. (1976). "Risk, return and arbitrage", in *Risk and Return in Finance* 2, ed. Irwin Friend and James S. Bicksler, Ballinger, Cambridge, MA.

Saito, T., T. Kariya and T. Otsu (1988). "A generalization of the principal component analysis", *Jour. Japan Statist. Soc.* 18, 187-193.

Schmalensee, R. and R.R. Trippi (1978). "Common stock volatility expectations implied by option premia", *Jour. Finance* 33, 1, 129-147.

Singleton, K.J. (1980). "A latent time series model of the cyclical behaviour of interest rates", *International Economic Review* 21, 559-575.

Siotani, M., T. Hayakawa and Y. Fujikoshi (1985). *Modern Multivariate Statistics: A graduate Course and Handbook*, American Sciences Press, Columbus.

Subba Rao T. and M.M. Gabr (1980). "A test for linearity of stationary time series", *Jour. Time Series Anal.* 1, 145-158.

Sugano, T., N. Mori, S. Tawada, T. Iwaishi and T. Yasuhisa (1990). An application of "Regression-MTV mixed model to Japanese stock prices" (in Japanese). Report at the Meeting of Japan Statistical Society.

Swamy, P.A.V.B. (1970). "Efficient inference in a random coefficient model", *Econometrica* 38, 311-323.

Takahashi, H. (1989). "Multifactor model and stock risk evaluation" (in Japanese), *Gendai Shoken Toshi Giho no Shintenkai*, Nihon Keizai Shinbun.

Takahashi, M. (1992). "Evaluation methods for derivative securities using the Hurst exponent" (in Japanese), Master thesis submitted to Tsukuba University.

Takano, T. (1989). "Stock evaluation model via multidimensional analysis" (in Japanese), *Gendai Shoken Toshi Giho no Shintenkai*, Nihon Keizai Shinbun.

Taqqu, M. (1975). "Weak convergence to fractional Brownian motion and to the Rosenblatt process", Z. Wahrscheinlichkeitstheorie, Gebiete 31, 287-302.

Tauchen, G. E. and M. Pitts (1983). "The price variability-volume relationship on speculative markets", Econometrica 51, 485-505.

Taylor, S. (1982). "Tests of the random walk hypothesis against a price-trend hypothesis", Jour. Financial and Quant. Anal. 17, 37-61.

Taylor, S. (1986). Modelling Financial Time Series, John Wiley, New York.

Tiao, G. C. and R. S. Tsay (1989). "Model specification in multi-variate time series", Jour. Roy. Statist. Soc. B51, 157-213.

Tiao, G. C. and Tsay, R. S. (1983). "Multiple time series modelling extended sample cross-correlations", Jour. Bus. Econ. Statist. 1, 43-56.

Tong, H. (1983). Threshold Models in Nonlinear Time Series Analysis, Springer-Velag, New York.

Tong, H. and K. S. Lim (1980). "Threshold autoregression, limit cycles and cyclical data", Jour. Roy. Statist. Soc. B42, 245-292.

Tsay, R. (1987). "Conditional heteroscedastic time series models", Jour. Amer. Statist. Assoc. 82, 590-604.

Tsay, R. S. (1989). "Testing and modelling threshold autoregressive processes", JASA 84, 231-240.

Tsay, R. S. (1988). "Nonlinear time series analysis: diagnostics and modelling", Statistical Sinica.

Tsuda, H. (1990). "An analysis on Japanese stock market via factor analysis model" (in Japanese) in Kinyu Shoken Keiryo Bunseki no Kiso to Oyo, Toyokeizai shinposha.

Tumura, Y. and K. Fukutomi (1970). "On the improper solutions in factor analysis", TRU Mathematics 6, 63-71.

Tumura, Y. and M. Sato (1981a). "On the covergence of iterative procedures in factor analysis", TRU Mathematics 17-1, 159-167.

Westerfield, J. M. (1977). "The distribution of common stock price changes; an application of transaction time and subordinated stochastic models", Jour. Financial and Quant. Anal. 10, 743-765.

Whaley, R. E. (1982). "Valuation of American call options on dividend-paying stocks. Empirical tests", Jour. Financial Econ. 10, 29-58.

Yajima, Y. (1985). "On estimation of long time series models", Austral. Jour. Statist. 27, 303-320.

Yamada, M. (1990). "Analysis on CAPM", (in Japanese) in *Kinyu Shoken Keiryo Bunseki no Kiso to Oyo*, Toyokeizai Shinposha.

Zellner, A. (1962). "An efficient method of estimating seemingly unrelated regression equations: some exact finite sample results", *Jour. Amer. Statist.* *Assoc.* **58**, 977–992.

INDEX

THEORY AND DECISION LIBRARY

SERIES B: MATHEMATICAL AND STATISTICAL METHODS
Editor: H. J. Skala, *University of Paderborn, Germany*

1. D. Rasch and M.L. Tiku (eds.): *Robustness of Statistical Methods and Nonparametric Statistics.* 1984   ISBN 90-277-2076-2

2. J.K. Sengupta: *Stochastic Optimization and Economic Models.* 1986
                                                               ISBN 90-277-2301-X

3. J. Aczél: *A Short Course on Functional Equations.* Based upon Recent Applications to the Social Behavioral Sciences. 1987
                                              ISBN Hb 90-277-2376-1; Pb 90-277-2377-X

4. J. Kacprzyk and S.A. Orlovski (eds.): *Optimization Models Using Fuzzy Sets and Possibility Theory.* 1987   ISBN 90-277-2492-X

5. A.K. Gupta (ed.): *Advances in Multivariate Statistical Analysis.* Pillai Memorial Volume. 1987   ISBN 90-277-2531-4

6. R. Kruse and K.D. Meyer: *Statistics with Vague Data.* 1987
                                                               ISBN 90-277-2562-4

7. J.K. Sengupta: *Applied Mathematics for Economics.* 1987
                                                               ISBN 90-277-2588-8

8. H. Bozdogan and A.K. Gupta (eds.): *Multivariate Statistical Modeling and Data Analysis.* 1987   ISBN 90-277-2592-6

9. B.R. Munier (ed.): *Risk, Decision and Rationality.* 1988
                                                               ISBN 90-277-2624-8

10. F. Seo and M. Sakawa: *Multiple Criteria Decision Analysis in Regional Planning.* Concepts, Methods and Applications. 1988   ISBN 90-277-2641-8

11. I. Vajda: *Theory of Statistical Inference and Information.* 1989
                                                               ISBN 90-277-2781-3

12. J.K. Sengupta: *Efficiency Analysis by Production Frontiers.* The Non-parametric Approach. 1989   ISBN 0-7923-0028-9

13. A. Chikán (ed.): *Progress in Decision, Utility and Risk Theory.* 1991
                                                               ISBN 0-7923-1211-2

14. S.E. Rodabaugh, E.P. Klement and U. Höhle (eds.): *Applications of Category Theory to Fuzzy Subsets.* 1992   ISBN 0-7923-1511-1

15. A. Rapoport: *Decision Theory and Decision Behaviour.* Normative and Descriptive Approaches. 1989   ISBN 0-7923-0297-4

16. A. Chikán (ed.): *Inventory Models.* 1990   ISBN 0-7923-0494-2

17. T. Bromek and E. Pleszczyńska (eds.): *Statistical Inference.* Theory and Practice. 1991   ISBN 0-7923-0718-6

18. J. Kacprzyk and M. Fedrizzi (eds.): *Multiperson Decision Making Models Using Fuzzy Sets and Possibility Theory.* 1990   ISBN 0-7923-0884-0

19. G.L. Gómez M.: *Dynamic Probabilistic Models and Social Structure.* Essays on Socioeconomic Continuity. 1992   ISBN 0-7923-1713-0

20. H. Bandemer and W. Näther: *Fuzzy Data Analysis.* 1992
ISBN 0-7923-1772-6

21. A.G. Sukharev: *Minimax Models in the Theory of Numerical Methods.* 1992
ISBN 0-7923-1821-8

22. J. Geweke (ed.): *Decision Making under Risk and Uncertainty.* New Models and Empirical Findings. 1992   ISBN 0-7923-1904-4

23. T. Kariya: *Quantitative Methods for Portfolio Analysis.* MTV Model Approach. 1993   ISBN 0-7923-2254-1

KLUWER ACADEMIC PUBLISHERS – DORDRECHT / BOSTON / LONDON

The manufacturer's authorised representative in the EU is Springer
Nature Customer Service Centre GmbH, Europaplatz 3, 69115 Heidelberg,
Germany. If you have any concerns regarding our products, please
contact ProductSafety@springernature.com

Printed and bound by CPI Group (UK) Ltd, Croydon, CR0 4YY

29/04/2026

02099527-0001